The Factors of Organic Evolution

By

Herbert Spencer

Theophania Publishing

PREFACE.

The two parts of which this Essay consists, originally published in *The Nineteenth Century* for April and May 1886 respectively, now reappear with the assent of the proprietor and editor of that periodical, to whom my thanks are due for his courtesy in giving it. Some passages of considerable length which, with a view to needful brevity, were omitted when the articles first appeared, have been restored.

Though the direct bearings of the arguments contained in this Essay are biological, the argument contained in its first half has indirect bearings upon Psychology, Ethics, and Sociology. My belief in the profound importance of these indirect bearings, was originally a chief prompter to set forth the argument; and it now prompts me to re-issue it in permanent form.

Though mental phenomena of many kinds, and especially of the simpler kinds, are explicable only as resulting from the natural selection of favourable variations; yet there are, I believe, still more numerous mental phenomena, including all those of any considerable complexity, which cannot be explained otherwise than as results of the inheritance of functionally-produced modifications. What theory of psychological evolution is espoused, thus depends on acceptance or rejection of the doctrine that not only in the individual, but in the successions of individuals, use and disuse of parts produce respectively increase and decrease of them.

Of course there are involved the conceptions we form of the genesis and nature of our higher emotions; and, by implication, the conceptions we form of our moral intuitions. If functionally-produced modifications are inheritable, then the mental associations habitually produced in individuals by experiences of the relations between actions and their consequences, pleasurable or painful, may, in the successions of individuals, generate innate tendencies to like or dislike such actions. But if not, the genesis of such tendencies is, as we shall see, not satisfactorily explicable.

That our sociological beliefs must also be profoundly affected by the conclusions we draw on this point, is obvious. If a nation is modified *en masse* by transmission of the effects produced on the natures of its members by those modes of daily activity which its institutions and circumstances involve; then we must infer that such institutions and circumstances mould its members far more rapidly and comprehensively than they can do if the sole cause of adaptation to them is the more frequent survival of individuals who happen to have varied in favourable ways.

I will add only that, considering the width and depth of the effects which acceptance of one or other of these hypotheses must have on our views of Life, Mind, Morals, and Politics, the question—Which of them is true? demands, beyond all other questions whatever, the attention of scientific men.

Brighton, January, 1887.

THE FACTORS OF ORGANIC EVOLUTION.

[*April and May, 1886*]

I.

Within the recollection of men now in middle life, opinion concerning the derivation of animals and plants was in a chaotic state. Among the unthinking there was tacit belief in creation by miracle, which formed an essential part of the creed of Christendom; and among the thinking there were two parties, each of which held an indefensible hypothesis. Immensely the larger of these parties, including nearly all whose scientific culture gave weight to their judgments, though not accepting literally the theologically-orthodox doctrine, made a compromise between that doctrine and the doctrines which geologists had established; while opposed to them were some, mostly having no authority in science, who held a doctrine which was heterodox both theologically and scientifically. Professor Huxley, in his lecture on "The Coming of Age of the Origin of Species," remarks concerning the first of these parties as follows:—

"One-and-twenty years ago, in spite of the work commenced by Hutton and continued with rare skill and patience by Lyell, the dominant view of the past history of the earth was catastrophic. Great and sudden physical revolutions, wholesale creations and extinctions of living beings, were the ordinary machinery of the geological epic brought into fashion by the misapplied genius of Cuvier. It was gravely maintained and taught that the end of every geological epoch was signalised by a cataclysm, by which every living being on the globe was swept away, to be replaced by a brand-new creation when the world returned to quiescence. A scheme of nature which appeared to be modelled on the likeness of a succession of rubbers of whist, at the end of each of which the players upset the table and called for a new pack, did not seem to shock anybody.

I may be wrong, but I doubt if, at the present time, there is a single responsible representative of these opinions left. The progress of scientific geology has elevated the fundamental principle of uniformitarianism, that the explanation of the past is to be sought in the study of the present, into the position of an axiom; and the wild speculations of the catastrophists, to which we all listened with respect a quarter of a century ago, would hardly find a single patient hearer at the present day."

Of the party above referred to as not satisfied with this conception described by Professor Huxley, there were two classes. The great majority were admirers of the *Vestiges of the Natural History of Creation*—a work which, while it sought to show that organic evolution has taken place, contended that the cause of organic evolution, is "an impulse" supernaturally "imparted to the forms of life, advancing them, ... through grades of organization." Being nearly all very inadequately acquainted with the facts, those who accepted the view set forth in the *Vestiges* were ridiculed by the well-instructed for being satisfied with evidence, much of which was either invalid or easily cancelled by counter-evidence, and at the same time they exposed themselves to the ridicule of the more philosophical for being content with a supposed explanation which was in reality no explanation: the alleged "impulse" to advance giving us no more help in understanding the facts than does Nature's alleged "abhorrence of a vacuum" help us to understand the ascent of water in a pump. The remnant, forming the second of these classes, was very small. While rejecting this mere verbal solution, which both Dr. Erasmus Darwin and Lamarck had shadowed forth in other language, there were some few who, rejecting also the hypothesis indicated by both Dr. Darwin and Lamarck, that the promptings of desires or wants produced growths of the parts subserving them, accepted the single *vera causa* assigned by these writers—the modification of structures resulting from modification of functions. They recognized as the sole process in organic development, the adaptation of parts and powers consequent on the effects of use and disuse—that continual moulding and re-moulding of organisms to suit their circumstances, which is brought about by direct converse with such circumstances.

But while this cause accepted by these few is a true cause, since unquestionably during the life of the individual organism changes of function produce changes of structure; and while it is a tenable hypothesis that changes of structure so produced are inheritable; yet it was manifest to those not prepossessed, that this cause cannot with reason be assigned for the greater part of the facts. Though in plants there are some characters which may not irrationally be ascribed to the direct effects of modified functions consequent on modified circumstances, yet the majority of the traits presented by plants are not to be thus explained. It is impossible that the thorns by which a briar is in large measure defended against browsing animals, can have been developed and moulded by the continuous exercise of their protective actions; for in the first place, the great majority of the thorns are never touched at all, and, in the second place, we have no ground whatever for supposing that those which are touched are thereby made to grow, and to take those shapes which render them efficient. Plants which are rendered uneatable by the thick woolly coatings of their leaves, cannot have had these coatings produced by any process of reaction against the action of enemies; for there is no imaginable reason why, if one part of a plant is eaten, the rest should thereafter begin to develop the hairs on its surface. By what direct effect of function on structure, can the shell of a nut have been evolved? Or how can those seeds which contain essential oils, rendering them unpalatable to birds, have been made to secrete such essential oils by these actions of birds which they restrain? Or how can the delicate plumes borne by some seeds, and giving the wind power to waft them to new stations, be due to any immediate influences of surrounding conditions? Clearly in these and in countless other cases, change of structure cannot have been directly caused by change of function. So is it with animals to a large extent, if not to the same extent. Though we have proof that by rough usage the dermal layer may be so excited as to produce a greatly thickened epidermal layer, sometimes quite horny; and though it is a feasible hypothesis that an effect of this kind persistently produced may be inherited; yet no such cause can explain the carapace of the turtle, the armour of the armadillo, or the imbricated covering of the manis. The skins of these animals are no more exposed to habitual hard usage than are those of animals covered by hair. The strange excrescences which distinguish the heads of the hornbills, cannot possibly have arisen from any reaction against the action of surrounding forces; for even were they clearly protective, there is no reason to suppose that the heads of these birds need protection more than the heads of other birds. If, led by the evidence that in animals the amount of covering is in some cases affected by the degree of exposure, it were admitted as imaginable that the development of feathers from preceding dermal growths had resulted from that extra nutrition caused by extra superficial circulation, we should still be without explanation of the structure of a feather. Nor should we have any clue to the specialities of feathers—the crests of various birds, the tails sometimes so enormous, the curiously placed plumes of the bird of paradise, &c., &c. Still more obviously impossible is it to explain as due to use or disuse the colours of animals. No direct adaptation to function could have produced the blue protuberances on a mandril's face, or the striped hide of a tiger, or the gorgeous plumage of a kingfisher, or the eyes in a peacock's tail, or the multitudinous patterns of insects' wings. One single case, that of a deer's horns, might alone have sufficed to show how insufficient was the assigned cause. During their growth, a deer's horns are not used at all; and when, having been cleared of the dead skin and dried-up blood-vessels covering them, they are ready for use, they are nerveless and non-vascular, and hence are incapable of undergoing any changes of structure consequent on changes of function.

Of these few then, who rejected the belief described by Professor Huxley, and who, espousing the belief in a continuous evolution, had to account for this evolution, it must be said that though the cause assigned was a true cause, yet, even admitting that it operated through successive generations, it left unexplained the greater part of the facts. Having been myself one of these few, I look back with surprise at the way in which the facts which were congruous with the espoused view monopolized consciousness and kept out the facts which were incongruous with it—conspicuous though many of them were. The misjudgment was not unnatural. Finding it impossible to accept any doctrine which implied a breach in the uniform course of natural causation, and, by implication, accepting as unquestionable the origin and development of all organic forms by accumulated modifications naturally caused, that which appeared to explain certain classes of these modifications, was supposed to be capable of explaining the rest: the tendency being to assume that these would eventually be similarly accounted for, though it was not clear how.

Returning from this parenthetic remark, we are concerned here chiefly to remember that, as said at the outset, there existed thirty years ago, no tenable theory about the genesis of living things. Of the two alternative beliefs, neither would bear critical examination.

Out of this dead lock we were released—in large measure, though not I believe entirely—by the *Origin of Species*. That work brought into view a further factor; or rather, such factor, recognized as in operation by here and there an observer (as pointed out by Mr. Darwin in his introduction to the second edition), was by him for the first time seen to have played so immense a part in the genesis of plants and animals.

Though laying myself open to the charge of telling a thrice-told tale, I feel obliged here to indicate briefly the several great classes of facts which Mr. Darwin's hypothesis explains; because otherwise that which follows would scarcely be understood. And I feel the less hesitation in doing this because the hypothesis which it replaced, not very widely known at any time, has of late so completely dropped into the background, that the majority of readers are scarcely aware of its existence, and do not therefore understand the relation between Mr. Darwin's successful interpretation and the preceding unsuccessful attempt at interpretation. Of these classes of facts, four chief ones may be here distinguished.

In the first place, such adjustments as those exemplified above are made comprehensible. Though it is inconceivable that a structure like that of the pitcher-plant could have been produced by accumulated effects of function on structure; yet it is conceivable that successive selections of favourable variations might have produced it; and the like holds of the no less remarkable appliance of the Venus's Fly-trap, or the still more astonishing one of that water-plant by which infant-fish are captured. Though it is impossible to imagine how, by direct influence of increased use, such dermal appendages as a porcupine's quills could have been developed; yet, profiting as the members of a species otherwise defenceless might do by the stiffness of their hairs, rendering them unpleasant morsels to eat, it is a feasible supposition that from successive survivals of individuals thus defended in the greatest degrees, and the consequent growth in successive generations of hairs into bristles, bristles into spines, spines into quills (for all these are homologous), this change could have arisen. In like manner, the odd inflatable bag of the bladder-nosed seal, the curious fishing-rod with its worm-like appendage carried on the head of the *lophius* or angler, the spurs on the wings of certain birds, the weapons of the sword-fish and saw-fish, the wattles of fowls, and numberless such peculiar structures, though by no possibility explicable as due to effects of use or disuse, are explicable as resulting from natural selection operating in one or other way.

In the second place, while showing us how there have arisen countless modifications in the forms, structures, and colours of each part, Mr. Darwin has shown us how, by the establishment of favourable variations, there may arise new parts. Though the first step in the production of horns on the heads of various herbivorous animals, may have been the growth of callosities consequent on the habit of butting—such callosities thus functionally initiated being afterwards developed in the most advantageous ways by selection; yet no explanation can be thus given of the sudden appearance of a duplicate set of horns, as occasionally happens in sheep: an addition which, where it proved beneficial, might readily be made a permanent trait by natural selection. Again, the modifications which follow use and disuse can by no possibility account for changes in the numbers of vertebræ; but after recognizing spontaneous, or rather fortuitous, variation as a factor, we can see that where an additional vertebra hence resulting (as in some pigeons) proves beneficial, survival of the fittest may make it a constant character; and there may, by further like additions, be produced extremely long strings of vertebræ, such as snakes show us. Similarly with the mammary glands. It is not an unreasonable supposition that by the effects of greater or less function, inherited through successive generations, these may be enlarged or diminished in size; but it is out of the question to allege such a cause for changes in their numbers. There is no imaginable explanation of these save the establishment by inheritance of spontaneous variations, such as are known to occur in the human race.

So too, in the third place, with certain alterations in the connections of parts. According to the greater or smaller demands made on this or that limb, the muscles moving it may be augmented or diminished in bulk; and, if there is inheritance of changes so wrought, the limb may, in course of generations, be rendered larger or smaller. But changes in the arrangements or attachments of muscles cannot be thus accounted for. It is found, especially at the extremities, that the relations of tendons to bones and to one another are not always the same. Variations in their modes of connection may occasionally prove advantageous, and may thus become established. Here again, then, we have a class of structural changes to which Mr. Darwin's hypothesis gives us the key, and to which there is no other key.

Once more there are the phenomena of mimicry. Perhaps in a more striking way than any others, these show how traits which seem inexplicable are explicable as due to the more frequent survival of individuals that have varied in favorable ways. We are enabled to understand such marvelous simulations as those of the leaf-insect, those of beetles which "resemble glittering dew-drops upon the leaves;" those of caterpillars which, when asleep, stretch themselves out so as to look like twigs. And we are shown how there have arisen still more astonishing imitations—those of one insect by another. As Mr. Bates has proved, there are cases in which a species of butterfly, rendered so unpalatable to insectivorous birds by its disagreeable taste that they will not catch it, is simulated in its colors and markings by a species which is structurally quite different—so simulated that even a practiced entomologist is liable to be deceived: the explanation being that an original slight resemblance, leading to occasional mistakes on the part of birds, was increased generation after generation by the more frequent escape of the most-like individuals, until the likeness became thus great.

But now, recognizing in full this process brought into clear view by Mr. Darwin, and traced out by him with so much care and skill, can we conclude that, taken alone, it accounts for organic evolution? Has the natural selection of favourable variations been the sole factor? On critically examining the evidence, we shall find reason to think that it by no means explains all that has to be explained. Omitting for the present any consideration of a factor which may be distinguished as primordial, it may be contended that the above-named factor alleged by Dr. Erasmus Darwin and by Lamarck, must be recognized as a co-operator. Utterly inadequate to explain the major part of the facts as is the hypothesis of the inheritance of functionally-produced modifications, yet there is a minor part of the facts, very extensive though less, which must be ascribed to this cause.

When discussing the question more than twenty years ago (*Principles of Biology*, § 166), I instanced the decreased size of the jaws in the civilized races of mankind, as a change not accounted for by the natural selection of favourable variations; since no one of the decrements by which, in thousands of years, this reduction has been effected, could have given to an individual in which it occurred, such advantage as would cause his survival, either through diminished cost of local nutrition or diminished weight to be carried. I did not then exclude, as I might have done, two other imaginable causes. It may be said that there is some organic correlation between increased size of brain and decreased size of jaw: Camper's doctrine of the facial angle being referred to in proof. But this argument may be met by pointing to the many examples of small-jawed people who are also small-brained, and by citing not infrequent cases of individuals remarkable for their mental powers, and at the same time distinguished by jaws not less than the average but greater. Again, if sexual selection be named as a possible cause, there is the reply that, even supposing such slight diminution of jaw as took place in a single generation to have been an attraction, yet the other incentives to choice on the part of men have been too many and great to allow this one to weigh in an adequate degree; while, during the greater portion of the period, choice on the part of women has scarcely operated: in earlier times they were stolen or bought, and in later times mostly coerced by parents. Thus, reconsideration of the facts does not show me the invalidity of the conclusion drawn, that this decrease in size of jaw can have had no other cause than continued inheritance of those diminutions consequent on diminutions of function, implied by the use of selected and well-prepared food. Here, however, my chief purpose is to add an instance showing, even more clearly, the connexion between change of function and change of structure. This instance, allied in nature to the other, is presented by those varieties, or rather sub-varieties, of dogs, which, having been household pets, and habitually fed on soft food, have not been called on to use their jaws in tearing and crunching, and have been but rarely allowed to use them in catching prey and in fighting. No inference can be drawn from the sizes of the jaws themselves, which, in these dogs, have probably been shortened mainly by selection. To get direct proof of the decrease of the muscles concerned in closing the jaws or biting, would require a series of observations very difficult to make. But it is not difficult to get indirect proof of this decrease by looking at the bony structures with which these muscles are connected. Examination of the skulls of sundry indoor dogs contained in the Museum of the College of Surgeons, proves the relative smallness of such parts. The only pug-dog's skull is that of an individual not perfectly adult; and though its traits are quite to the point they cannot with safety be taken as evidence. The skull of a toy-terrier has much restricted areas of insertion for the temporal muscles; has weak zygomatic arches; and has extremely small attachments for the masseter muscles. Still more significant is the evidence furnished by the skull of a King Charles's spaniel, which, if we allow three years to a generation, and bear in mind that the variety must have existed before Charles the Second's reign, we may assume belongs to something approaching to the hundredth generation of these household pets. The relative breadth between the outer surfaces of the zygomatic arches is conspicuously small; the narrowness of the temporal fossæ is also striking; the zygomata are very slender; the temporal muscles have left no marks whatever, either by limiting lines or by the character of the surfaces covered; and the places of attachment for the masseter muscles are very feebly developed. At the Museum of Natural History, among skulls of dogs there is one which, though unnamed, is shown by its small size and by its teeth, to have belonged to one variety or other of lap-dogs, and which has the same traits in an equal degree with the skull just described. Here, then, we have two if not three kinds of dogs which, similarly leading protected and pampered lives, show that in the course of generations the parts concerned in clenching the jaws have dwindled. To what cause must this decrease be ascribed? Certainly not to artificial selection; for most of the modifications named make no appreciable external signs: the width across the zygomata could alone be perceived. Neither can natural selection have had anything to do with it; for even were there any struggle for existence among such dogs, it cannot be contended that any advantage in the struggle could be gained by an individual in which a decrease took place. Economy of nutrition, too, is excluded. Abundantly fed as such dogs are, the constitutional tendency is to find places where excess of absorbed nutriment may be conveniently deposited, rather than to find places where some cutting down of the supplies is practicable. Nor again can there be alleged a possible correlation between these diminutions and that shortening of the jaws which has probably resulted from selection; for in the bull-dog, which has also relatively short jaws, these structures concerned in closing them are unusually large. Thus there remains as the only conceivable cause, the diminution of size which results from diminished use. The dwindling of a little-exercised part has, by inheritance, been made more and more marked in successive generations.

Difficulties of another class may next be exemplified—those which present themselves when we ask how there can be effected by the selection of favourable variations, such changes of structure as adapt an organism to some useful action in which many different parts co-operate. None can fail to see how a simple part may, in course of generations, be greatly enlarged, if each enlargement furthers, in some decided way, maintenance of the species. It is easy to understand, too, how a complex part, as an entire limb, may be increased as a whole by the simultaneous due increase of its co-operative parts; since if, while it is growing, the channels of supply bring to the limb an unusual quantity of blood, there will naturally result a proportionately greater size of all its components—bones, muscles, arteries, veins, &c. But though in cases like this, the co-operative parts forming some large complex part may be expected to vary together, nothing implies that they necessarily do so; and we have proof that in various cases, even when closely united, they do not do so. An example is furnished by those blind crabs named in the *Origin of Species* which inhabit certain dark caves of Kentucky, and which, though they have lost their eyes, have not lost the foot-stalks which carried their eyes. In describing the varieties which have been produced by pigeon-fanciers, Mr. Darwin notes the fact that along with changes in length of beak produced by selection, there have not gone proportionate changes in length of tongue. Take again the case of teeth and jaws. In mankind these have not varied together. During civilization the jaws have decreased, but the teeth have not decreased in proportion; and hence that prevalent crowding of them, often remedied in childhood by extraction of some, and in other cases causing that imperfect development which is followed by early decay. But the absence of proportionate variation in co-operative parts that are close together, and are even bound up in the same mass, is best seen in those varieties of dogs named above as illustrating the inherited effects of disuse. We see in them, as we see in the human race, that diminution in the jaws has not been accompanied by corresponding diminution in the teeth. In the catalogue of the College of Surgeons Museum, there is appended to the entry which identifies a Blenheim Spaniel's skull, the words—"the teeth are closely crowded together," and to the entry concerning the skull of a King Charles's Spaniel the words—"the teeth are closely packed, p. 3, is placed quite transversely to the axis of the skull." It is further noteworthy that in a case where there is no diminished use of the jaws, but where they have been shortened by selection, a like want of concomitant variation is manifested: the case being that of the bull-dog, in the upper jaw of which also, "the premolars ... are excessively crowded, and placed obliquely or even transversely to the long axis of the skull."[1]

If, then, in cases where we can test it, we find no concomitant variation in co-operative parts that are near together—if we do not find it in parts which, though belonging to different tissues, are so closely united as teeth and jaws—if we do not find it even when the co-operative parts are not only closely united, but are formed out of the same tissue, like the crab's eye and its peduncle; what shall we say of co-operative parts which, besides being composed of different tissues, are remote from one another? Not only are we forbidden to assume that they vary together, but we are warranted in asserting that they can have no tendency to vary together. And what are

---

[1] 1. It is probable that this shortening has resulted not directly but indirectly, from the selection of individuals which were noted for tenacity of hold; for the bull-dog's peculiarity in this respect seems due to relative shortness of the upper jaw, giving the underhung structure which, involving retreat of the nostrils, enables the dog to continue breathing while holding.

the implications in cases where increase of a structure can be of no service unless there is concomitant increase in many distant structures, which have to join it in performing the action for which it is useful?

As far back as 1864 (*Principles of Biology*, § 166) I named in illustration an animal carrying heavy horns—the extinct Irish elk; and indicated the many changes in bones, muscles, blood-vessels, nerves, composing the fore-part of the body, which would be required to make an increment of size in such horns advantageous. Here let me take another instance— that of the giraffe: an instance which I take partly because, in the sixth edition of the *Origin of Species*, issued in 1872, Mr. Darwin has referred to this animal when effectually disposing of certain arguments urged against his hypothesis. He there says:—

> "In order that an animal should acquire some structure specially and largely developed, it is almost indispensable that several other parts should be modified and co-adapted. Although every part of the body varies slightly, it does not follow that the necessary parts should always vary in the right direction and to the right degree" (p. 179).

And in the summary of the chapter, he remarks concerning the adjustments in the same quadruped, that "the prolonged use of all the parts together with inheritance will have aided in an important manner in their co-ordination" (p. 199): a remark probably having reference chiefly to the increased massiveness of the lower part of the neck; the increased size and strength of the thorax required to bear the additional burden; and the increased strength of the fore-legs required to carry the greater weight of both. But now I think that further consideration suggests the belief that the entailed modifications are much more numerous and remote than at first appears; and that the greater part of these are such as cannot be ascribed in any degree to the selection of favourable variations, but must be ascribed exclusively to the inherited effects of changed functions. Whoever has seen a giraffe gallop will long remember the sight as a ludicrous one. The reason for the strangeness of the motions is obvious. Though the fore limbs and the hind limbs differ so much in length, yet in galloping they have to keep pace—must take equal strides. The result is that at each stride, the angle which the hind limbs describe round their centre of motion is much larger than the angle described by the fore limbs. And beyond this, as an aid in equalizing the strides, the hind part of the back is at each stride bent very much downwards and forwards. Hence the hind-quarters appear to be doing nearly all the work. Now a moment's observation shows that the bones and muscles composing the hind-quarters of the giraffe, perform actions differing in one or other way and degree, from the actions performed by the homologous bones and muscles in a mammal of ordinary proportions, and from those in the ancestral mammal which gave origin to the giraffe. Each further stage of that growth which produced the large fore-quarters and neck, entailed some adapted change in sundry of the numerous parts composing the hind-quarters; since any failure in the adjustment of their respective strengths would entail some defect in speed and consequent loss of life when chased. It needs but to remember how, when continuing to walk with a blistered foot, the taking of steps in such a modified way as to diminish pressure on the sore point, soon produces aching of muscles which are called into unusual action, to see that over-straining of any one of the muscles of the giraffe's hind-quarters might quickly incapacitate the animal when putting out all its powers to escape; and to be a few yards behind others would cause death. Hence if we are debarred from assuming that co-operative parts vary together even when adjacent and closely united—if we are still more debarred from assuming that with increased length of fore-legs or of neck, there will go an appropriate change in any one muscle or bone in the hind-quarters; how entirely out of the question it is to assume that there will simultaneously take place the appropriate changes in *all* those many components of the hind-quarters which severally require re-adjustment. It is useless to reply that an increment of length in the fore-legs or neck might be retained and transmitted to posterity, waiting an appropriate variation in a particular bone or muscle in the hind-quarters, which, being made, would allow of a further increment. For besides the fact that until this secondary variation occurred the primary variation would be a disadvantage often fatal; and besides the fact that before such an appropriate secondary variation might be expected in the course of generations to occur, the primary variation would have died out; there is the fact that the appropriate variation of one bone or muscle in the hind-quarters would be useless without appropriate variations of all the rest—some in this way and some in that—a number of appropriate variations which it is impossible to suppose.

Nor is this all. Far more numerous appropriate variations would be indirectly necessitated. The immense change in the ratio of fore-quarters to hind-quarters would make requisite a corresponding change of ratio in the appliances carrying on the nutrition of the two. The entire vascular system, arterial and venous, would have to undergo successive unbuildings and rebuildings to make its channels everywhere adequate to the local requirements; since any want of adjustment in the blood-supply in this or that set of muscles, would entail incapacity, failure of speed, and loss of life. Moreover the nerves supplying the various sets of muscles would have to be proportionately changed; as well as the central nervous tracts from which they issued. Can we suppose that all these appropriate changes, too, would be step by step simultaneously made by fortunate spontaneous variations, occurring along with all the other fortunate spontaneous variations? Considering how immense must be the number of these required changes, added to the changes above enumerated, the chances against any adequate re-adjustments fortuitously arising must be infinity to one.

If the effects of use and disuse of parts are inheritable, then any change in the fore parts of the giraffe which affects the action of the hind limbs and back, will simultaneously cause, by the greater or less exercise of it, a re-moulding of each component in the hind limbs and back in a way adapted to the new demands; and generation after generation the entire structure of the hind-quarters will be progressively fitted to the changed structure of the fore-quarters: all the appliances for nutrition and innervation being at the same time progressively fitted to both. But in the absence of this inheritance of functionally-produced modifications, there is no seeing how the required readjustments can be made.

Yet a third class of difficulties stands in the way of the belief that the natural selection of useful variations is the sole factor of organic evolution. This class of difficulties, already pointed out in § 166 of the *Principles of Biology*, I cannot more clearly set forth than in the words there used. Hence I may perhaps be excused for here quoting them.

> "Where the life is comparatively simple, or where surrounding circumstances render some one function supremely important, the survival of the fittest may readily bring about the appropriate structural change, without any aid from the transmission of functionally-acquired modifications. But in proportion as the life grows complex—in proportion as a healthy existence cannot be secured by a large endowment of some one power, but demands many powers; in the same proportion do there arise obstacles to the increase of any particular power, by "the preservation of favoured races in the struggle for life." As fast as the faculties are multiplied, so fast does it become possible

for the several members of a species to have various kinds of superiorities over one another. While one saves its life by higher speed, another does the like by clearer vision, another by keener scent, another by quicker hearing, another by greater strength, another by unusual power of enduring cold or hunger, another by special sagacity, another by special timidity, another by special courage; and others by other bodily and mental attributes. Now it is unquestionably true that, other things equal, each of these attributes, giving its possessor an extra chance of life, is likely to be transmitted to posterity. But there seems no reason to suppose that it will be increased in subsequent generations by natural selection. That it may be thus increased, the individuals not possessing more than average endowments of it, must be more frequently killed off than individuals highly endowed with it; and this can happen only when the attribute is one of greater importance, for the time being, than most of the other attributes. If those members of the species which have but ordinary shares of it, nevertheless survive by virtue of other superiorities which they severally possess; then it is not easy to see how this particular attribute can be developed by natural selection in subsequent generations. The probability seems rather to be, that by gamogenesis, this extra endowment will, on the average, be diminished in posterity—just serving in the long run to compensate the deficient endowments of other individuals, whose special powers lie in other directions; and so to keep up the normal structure of the species. The working out of the process is here somewhat difficult to follow; but it appears to me that as fast as the number of bodily and mental faculties increases, and as fast as the maintenance of life comes to depend less on the amount of any one, and more on the combined action of all; so fast does the production of specialities of character by natural selection alone, become difficult. Particularly does this seem to be so with a species so multitudinous in its powers as mankind; and above all does it seem to be so with such of the human powers as have but minor shares in aiding the struggle for life—the æsthetic faculties, for example."

Dwelling for a moment on this last illustration of the class of difficulties described, let us ask how we are to interpret the development of the musical faculty. I will not enlarge on the family antecedents of the great composers. I will merely suggest the inquiry whether the greater powers possessed by Beethoven and Mozart, by Weber and Rossini, than by their fathers, were not due in larger measure to the inherited effects of daily exercise of the musical faculty by their fathers, than to inheritance, with increase, of spontaneous variations; and whether the diffused musical powers of the Bach clan, culminating in those of Johann Sebastian, did not result in part from constant practice; but I will raise the more general question— How came there that endowment of musical faculty which characterizes modern Europeans at large, as compared with their remote ancestors. The monotonous chants of low savages cannot be said to show any melodic inspiration; and it is not evident that an individual savage who had a little more musical perception than the rest, would derive any such advantage in the maintenance of life as would secure the spread of his superiority by inheritance of the variation. And then what are we to say of harmony? We cannot suppose that the appreciation of this, which is relatively modern, can have arisen by descent from the men in whom successive variations increased the appreciation of it—the composers and musical performers; for on the whole, these have been men whose worldly prosperity was not such as enabled them to rear many children inheriting their special traits. Even if we count the illegitimate ones, the survivors of these added to the survivors of the legitimate ones, can hardly be held to have yielded more than average numbers of descendants; and those who inherited their special traits have not often been thereby so aided in the struggle for existence as to further the spread of such traits. Rather the tendency seems to have been the reverse.

Since the above passage was written, I have found in the second volume of *Animals and Plants under Domestication*, a remark made by Mr. Darwin, practically implying that among creatures which depend for their lives on the efficiency of numerous powers, the increase of any one by the natural selection of a variation is necessarily difficult. Here it is.

> "Finally, as indefinite and almost illimitable variability is the usual result of domestication and cultivation, with the same part or organ varying in different individuals in different or even in directly opposite ways; and as the same variation, if strongly pronounced, usually recurs only after long intervals of time, any particular

variation would generally be lost by crossing, reversion, and the accidental destruction of the varying individuals, unless carefully preserved by man."—Vol. ii, 292.

Remembering that mankind, subject as they are to this domestication and cultivation, are not, like domesticated animals, under an agency which picks out and preserves particular variations; it results that there must usually be among them, under the influence of natural selection alone, a continual disappearance of any useful variations of particular faculties which may arise. Only in cases of variations which are specially preservative, as for example, great cunning during a relatively barbarous state, can we expect increase from natural selection alone. We cannot suppose that minor traits, exemplified among others by the æsthetic perceptions, can have been evolved by natural selection. But if there is inheritance of functionally-produced modifications of structure, evolution of such minor traits is no longer inexplicable.

Two remarks made by Mr. Darwin have implications from which the same general conclusion must, I think, be drawn. Speaking of the variability of animals and plants under domestication, he says:—

> "Changes of any kind in the conditions of life, even extremely slight changes, often suffice to cause variability.... Animals and plants continue to be variable for an immense period after their first domestication; ... In the course of time they can be habituated to certain changes, so as to become less variable; ... There is good evidence that the power of changed conditions accumulates; so that two, three, or more generations must be exposed to new conditions before any effect is visible.... Some variations are induced by the direct action of the surrounding conditions on the whole organization, or on certain parts alone, and other variations are induced indirectly through the reproductive system being affected in the same manner as is so common with organic beings when removed from their natural conditions."— (*Animals and Plants under Domestication*, vol. ii, 270.)

There are to be recognized two modes of this effect produced by changed conditions on the reproductive system, and consequently on offspring. Simple arrest of development is one. But beyond the variations of offspring arising from imperfectly developed reproductive systems in parents—variations which must be ordinarily in the nature of imperfections—there are others due to a changed balance of functions caused by changed conditions. The fact noted by Mr. Darwin in the above passage, "that the power of changed conditions accumulates; so that two, three, or more generations must be exposed to new conditions before any effect is visible," implies that during these generations there is going on some change of constitution consequent on the changed proportions and relations of the functions. I will not dwell on the implication, which seems tolerably clear, that this change must consist of such modifications of organs as adapt them to their changed functions; and that if the influence of changed conditions "accumulates," it must be through the inheritance of such modifications. Nor will I press the question—What is the nature of the effect registered in the reproductive elements, and which is subsequently manifested by variations?—Is it an effect entirely irrelevant to the new requirements of the variety?—Or is it an effect which makes the variety less fit for the new requirements?—Or is it an effect which makes it more fit for the new requirements? But not pressing these questions, it suffices to point out the necessary implication that changed functions of organs *do*, in some way or other, register themselves in changed proclivities of the reproductive elements. In face of these facts it cannot be denied that the modified action of a part produces an inheritable effect—be the nature of that effect what it may.

The second of the remarks above adverted to as made by Mr. Darwin, is contained in his sections dealing with correlated variations. In the *Origin of Species*, p. 114, he says—

> "The whole organization is so tied together during its growth and development, that when slight variations in any one part occur, and are accumulated through natural selection, other parts become modified."

And a parallel statement contained in *Animals and Plants under Domestication*, vol. ii, p. 320, runs thus—

> "Correlated variation is an important subject for us; for when one part is modified through continued selection, either by man or under nature, other parts of the organization will be unavoidably modified. From this correlation it apparently follows that, with our domesticated animals and plants, varieties rarely or never differ from each other by some single character alone."

By what process does a changed part modify other parts? By modifying their functions in some way or degree, seems the necessary answer. It is indeed, imaginable, that where the part changed is some dermal appendage which, becoming larger, has abstracted more of the needful material from the general stock, the effect may consist simply in diminishing the amount of this material available for other dermal appendages, leading to diminution of some or all of them, and may fail to affect in appreciable ways the rest of the organism: save perhaps the blood-vessels near the enlarged appendage. But where the part is an active one—a limb, or viscus, or any organ which constantly demands blood, produces waste matter, secretes, or absorbs—then all the other active organs become implicated in the change. The functions performed by them have to constitute a moving equilibrium; and the function of one cannot, by alteration of the structure performing it, be modified in degree or kind, without modifying the functions of the rest— some appreciably and others inappreciably, according to the directness or indirectness of their relations. Of such inter-dependent changes, the normal ones are naturally inconspicuous; but those which are partially or completely abnormal, sufficiently carry home the general truth. Thus, unusual cerebral excitement affects the excretion through the kidneys in quantity or quality or both. Strong emotions of disagreeable kinds check or arrest the flow of bile. A considerable obstacle to the circulation offered by some important structure in a diseased or disordered state, throwing more strain upon the heart, causes hypertrophy of its muscular walls; and this change which is, so far as concerns the primary evil, a remedial one, often entails mischiefs in other organs. "Apoplexy and palsy, in a scarcely credible number of cases, are directly dependent on hypertrophic enlargement of the heart." And in other cases, asthma, dropsy, and epilepsy are caused. Now if a result of this inter-dependence as seen in the individual organism, is that a local modification of one part produces, by changing their functions, correlative modifications of other parts, then the question here to be put is—Are these correlative modifications, when of a kind falling within normal limits, inheritable or not. If they are inheritable, then the fact stated by Mr. Darwin that "when one part is modified through continued selection," "other parts of the organization will be unavoidably modified" is perfectly intelligible: these entailed secondary modifications are transmitted *pari passu* with the successive modifications produced by selection. But what if they are not inheritable? Then these secondary modifications caused in the individual, not being transmitted to descendants, the descendants must commence life with organizations out of balance, and with each increment of change in the part affected by selection, their organizations must get more out of balance—must have a larger and larger amounts of re-organization to be made during their lives. Hence the constitution of the variety must become more and more unworkable.

The only imaginable alternative is that the re-adjustments are effected in course of time by natural selection. But, in the first place, as we find no proof of concomitant variation among directly co-operative parts which are closely united, there cannot be assumed any concomitant variation among parts which are both indirectly co-operative and far from one another. And, in the second place, before all the many required re-adjustments could be made, the variety would die out from defective constitution. Even were there no such difficulty, we should still have to entertain a strange group of propositions, which would stand as follows:—1. Change in one part entails, by reaction on the organism, changes, in other parts, the functions of which are necessarily changed. 2. Such changes worked in the individual, affect, in some way, the reproductive elements: these being found to evolve unusual structures when the constitutional balance has been continuously disturbed. 3. But the changes in the reproductive elements thus caused, are not such as represent these functionally-produced changes: the modifications conveyed to offspring are irrelevant to these various modifications functionally produced in the organs of the parents. 4. Nevertheless, while the balance of functions cannot be re-established through inheritance of the effects of disturbed functions on structures, wrought throughout the individual organism; it can be re-established by the inheritance of fortuitous variations which occur in all the affected organs without reference to these changes of function.

Now without saying that acceptance of this group of propositions is impossible, we may certainly say that it is not easy.

"But where are the direct proofs that inheritance of functionally-produced modifications takes place?" is a question which will be put by those who have committed themselves to the current exclusive interpretation. "Grant that there are difficulties; still, before the transmitted effects of use and disuse can be legitimately assigned in explanation of them, we must have good evidence that the effects of use and disuse *are* transmitted."

Before dealing directly with this demurrer, let me deal with it indirectly, by pointing out that the lack of recognized evidence may be accounted for without assuming that there is not plenty of it. Inattention and reluctant attention lead to the ignoring of facts which really exist in abundance; as is well illustrated in the case of pre-historic implements. Biassed by the current belief that no traces of man were to be found on the Earth's surface, save in certain superficial formations of very recent date, geologists and anthropologists not only neglected to seek such traces, but for a long time continued to pooh-pooh those who said they had found them. When M. Boucher de Perthes at length succeeded in drawing the eyes of scientific men to the flint implements discovered by him in the quarternary deposits of the Somme valley; and when geologists and anthropologists had thus been convinced that evidences of human existence were to be found in formations of considerable age, and thereafter began to search for them; they found plenty of them all over the world. Or again, to take an instance closely germane to the matter, we may recall the fact that the contemptuous attitude towards the hypothesis of organic evolution which naturalists in general maintained before the publication of Mr. Darwin's work, prevented them from seeing the multitudinous facts by which it is supported. Similarly, it is very possible that their alienation from the belief that there is a transmission of those changes of structure which are produced by changes of action, makes naturalists slight the evidence which supports that belief and refuse to occupy themselves in seeking further evidence.

If it be asked how it happens that there have been recorded multitudinous instances of variations fortuitously arising and re-appearing in offspring, while there have not been recorded instances of the transmission of changes functionally produced, there are three replies. The first is that changes of the one class are many of them conspicuous, while those of the other class are nearly all inconspicuous. If a child is born with six fingers, the anomaly is not simply obvious but so startling as to attract much notice; and if this child, growing up, has six-fingered descendants, everybody in the locality hears of it. A pigeon with specially-coloured feathers, or one distinguished by a broadened and upraised tail, or by a protuberance of the neck, draws attention by its oddness; and if in its young the trait is repeated, occasionally with increase, the fact is remarked, and there follows the thought of establishing the peculiarity by selection. A lamb disabled from leaping by the shortness of its legs, could not fail to be observed; and the fact that its offspring were similarly short-legged, and had a consequent inability to get over fences, would inevitably become widely known. Similarly with plants. That this flower had an extra number of petals, that that was unusually symmetrical, and that another differed considerably in colour from the average of its kind, would be easily seen by an observant gardener; and the suspicion that such anomalies are inheritable having arisen, experiments leading to further proofs that they are so, would frequently be made. But it is not thus with functionally-produced modifications. The seats of these are in nearly all cases the muscular, osseous, and nervous systems, and the viscera—parts which are either entirely hidden or greatly obscured. Modification in a nervous centre is inaccessible to vision; bones may be considerably altered in size or shape without attention being drawn to them; and, covered with thick coats as are most of the animals open to continuous observation, the increases or decreases in muscles must be great before they become externally perceptible.

A further important difference between the two inquiries is that to ascertain whether a fortuitous variation is inheritable, needs merely a little attention to the selection of individuals and the observation of offspring; while to ascertain whether there is inheritance of a functionally-produced modification, it is requisite to make arrangements which demand the greater or smaller exercise of some part or parts; and it is difficult in many cases to find such arrangements, troublesome to maintain them even for one generation, and still more through successive generations.

Nor is this all. There exist stimuli to inquiry in the one case which do not exist in the other. The money-interest and the interest of the fancier, acting now separately and now together, have prompted multitudinous individuals to make experiments which have brought out clear evidence that fortuitous variations are inherited. The cattle-breeders who profit by producing certain shapes and qualities; the keepers of pet animals who take pride in the perfections of those they have bred; the florists, professional and amateur, who obtain new varieties and take prizes; form a body of men who furnish naturalists with countless of the required proofs. But there is no such body of men, led either by pecuniary interest or the interest of a hobby, to ascertain by experiments whether the effects of use and disuse are inheritable.

Thus, then, there are amply sufficient reasons why there is a great deal of direct evidence in the one case and but little in the other: such little being that which comes out incidentally. Let us look at what there is of it.

Considerable weight attaches to a fact which Brown-Séquard discovered, quite by accident, in the course of his researches. He found that certain artificially-produced lesions of the nervous system, so small even as a section of the sciatic nerve, left, after healing, an increasing excitability which ended in liability to epilepsy; and there afterwards came out the unlooked-for result that the offspring of guinea-pigs which had thus acquired an epileptic habit such that a pinch on the neck would produce a fit, inherited an epileptic habit of like kind. It has, indeed, been since alleged that guinea pigs tend to epilepsy, and that phenomena of the kind described, occur where there have been no antecedents like those in Brown-Séquard's case. But considering the improbability that the phenomena observed by him happened to be nothing more than phenomena which occasionally arise naturally, we may, until there is good proof to the contrary, assign some value to his results.

Evidence not of this directly experimental kind, but nevertheless of considerable weight, is furnished by other nervous disorders. There is proof enough that insanity admits of being induced by circumstances which, in one or other way, derange the nervous functions—excesses of this or that kind; and no one questions the accepted belief that insanity is inheritable. Is it alleged that the insanity which is inheritable is that which spontaneously arises, and that the insanity which follows some chronic perversion of functions is not inheritable? This does not seem a very reasonable allegation; and until some warrant for it is forthcoming, we may fairly assume that there is here a further support for belief in the transmission of functionally-produced changes.

Moreover, I find among physicians the belief that nervous disorders of a less severe kind are inheritable. Men who have prostrated their nervous systems by prolonged overwork or in some other way, have children more or less prone to nervousness. It matters not what may be the form of inheritance—whether it be of a brain in some way imperfect, or of a deficient blood-supply; it is in any case the inheritance of functionally-modified structures.

Verification of the reasons above given for the paucity of this direct evidence, is yielded by contemplation of it; for it is observable that the cases named are cases which, from one or other cause, have thrust themselves on observation. They justify the suspicion that it is not because such cases are rare that many of them cannot be cited; but simply because they are mostly unobtrusive, and to be found only by that deliberate search which nobody makes. I say nobody, but I am wrong. Successful search has been made by one whose competence as an observer is beyond question, and whose testimony is less liable than that of all others to any bias towards the conclusion that such inheritance takes place. I refer to the author of the *Origin of Species*.

Now-a-days most naturalists are more Darwinian than Mr. Darwin himself. I do not mean that their beliefs in organic evolution are more decided; though I shall be supposed to mean this by the mass of readers, who identify Mr. Darwin's great contribution to the theory of organic evolution, with the theory of organic evolution itself, and even with the theory of evolution at large. But I mean that the particular factor which he first recognized as having played so immense a part in organic evolution, has come to be regarded by his followers as the sole factor, though it was not so regarded by him. It is true that he apparently rejected altogether the causal agencies alleged by earlier inquirers. In the Historical Sketch prefixed to the later editions of his *Origin of Species* (p. xiv, note), he writes:—"It is curious how largely my grandfather, Dr. Erasmus Darwin, anticipated the views and erroneous grounds of opinion of Lamarck in his 'Zoonomia' (vol. i, pp. 500-510), published in 1794." And since, among the views thus referred to, was the view that changes of structure in organisms arise by the inheritance of functionally-produced changes, Mr. Darwin seems, by the above sentence, to have implied his disbelief in such inheritance. But he did not mean to imply this; for his belief in it as a cause of evolution, if not an important cause, is proved by many passages in his works. In the first chapter of the *Origin of Species* (p. 11 of the first edition), he says respecting the inherited effects of habit, that "with animals the increased use or disuse of parts has had a marked influence;" and he gives as instances the changed relative weights of the wing bones and leg bones of the wild duck and the domestic duck, "the great and inherited development of the udders in cows and goats," and the drooping ears of various domestic animals. Here are other passages taken from the latest edition of the work.

> "I think there can be no doubt that use in our domestic animals has strengthened and enlarged certain parts, and disuse diminished them; and that such modifications are inherited" (p. 108). [And on the following pages he gives five further examples of such effects.] "Habit in producing constitutional peculiarities and use in strengthening and disuse in weakening and diminishing organs, appear in many cases to have been potent in their effects" (p. 131). "When discussing special

cases, Mr. Mivart passes over the effects of the increased use and disuse of parts, which I have always maintained to be highly important, and have treated in my 'Variation under Domestication' at greater length than, as I believe, any other writer" (p. 176). "Disuse, on the other hand, will account for the less developed condition of the whole inferior half of the body, including the lateral fins" (p. 188). "I may give another instance of a structure which apparently owes its origin exclusively to use or habit" (p. 188). "It appears probable that disuse has been the main agent in rendering organs rudimentary" (pp. 400-401). "On the whole, we may conclude that habit, or use and disuse, have, in some cases, played a considerable part in the modification of the constitution and structure; but that the effects have often been largely combined with, and sometimes overmastered by, the natural selection of innate variations" (p. 114).

In his subsequent work, *The Variation of Animals and Plants under Domestication*, where he goes into full detail, Mr. Darwin gives more numerous illustrations of the inherited effects of use and disuse. The following are some of the cases, quoted from volume i of the first edition.

> Treating of domesticated rabbits, he says:—"the want of exercise has apparently modified the proportional length of the limbs in comparison with the body" (p. 116). "We thus see that the most important and complicated organ [the brain] in the whole organization is subject to the law of decrease in size from disuse" (p. 129). He remarks that in birds of the oceanic islands "not persecuted by any enemies, the reduction of their wings has probably been caused by gradual disuse." After comparing one of these, the water-hen of Tristan d'Acunha, with the European water-hen, and showing that all the bones concerned in flight are smaller, he adds—"Hence in the skeleton of this natural species nearly the same changes have occurred, only carried a little further, as with our domestic ducks, and in this latter case I presume no one will dispute that they have resulted from the lessened use of the wings and the increased use of the legs" (pp. 286-7). "As with other long-domesticated animals, the instincts of the silk-moth have suffered.

> The caterpillars, when placed on a mulberry-tree, often commit the strange mistake of devouring the base of the leaf on which they are feeding, and consequently fall down; but they are capable, according to M. Robinet, of again crawling up the trunk. Even this capacity sometimes fails, for M. Martins placed some caterpillars on a tree, and those which fell were not able to remount and perished of hunger; they were even incapable of passing from leaf to leaf" (p. 304).

Here are some instances of like meaning from volume ii.

> "In many cases there is reason to believe that the lessened use of various organs has affected the corresponding parts in the offspring. But there is no good evidence that this ever follows in the course of a single generation.... Our domestic fowls, ducks, and geese have almost lost, not only in the individual but in the race, their power of flight; for we do not see a chicken, when frightened, take flight like a young pheasant.... With domestic pigeons, the length of the sternum, the prominence of its crest, the length of the scapulæ and furcula,

the length of the wings as measured from tip to tip of the radius, are all reduced relatively to the same parts in the wild pigeon." [After detailing kindred diminutions in fowls and ducks, Mr. Darwin adds] "The decreased weight and size of the bones, in the foregoing cases, is probably the indirect result of the reaction of the weakened muscles on the bones" (pp. 297-8). "Nathusius has shown that, with the improved races of the pig, the shortened legs and snout, the form of the articular condyles of the occiput, and the position of the jaws with the upper canine teeth projecting in a most anomalous manner in front of the lower canines, may be attributed to these parts not having been fully exercised.... These modifications of structure, which are all strictly inherited, characterise several improved breeds, so that they cannot have been derived from any single domestic or wild stock. With respect to cattle, Professor Tanner has remarked that the lungs and liver in the improved breeds 'are found to be considerably reduced in size when compared with those possessed by animals having perfect liberty....' The cause of the reduced lungs in highly-bred animals which take little exercise is obvious" (pp. 299-300). [And on pp. 301, 302 and 303, he gives facts showing the effects of use and disuse in changing, among domestic animals, the characters of the ears, the lengths of the intestines, and, in various ways, the natures of the instincts.]

But Mr. Darwin's admission, or rather his assertion, that the inheritance of functionally-produced modifications has been a factor in organic evolution, is made clear not by these passages alone and by kindred ones. It is made clearer still by a passage in the preface to the second edition of his *Descent of Man*. He there protests against that current version of his views in which this factor makes no appearance. The passage is as follows.

"I may take this opportunity of remarking that my critics frequently assume that I attribute all changes of corporeal structure and mental power exclusively to the natural selection of such variations as are often called spontaneous; whereas, even in the first edition of the 'Origin of Species,' I distinctly stated that great weight must be attributed to the inherited effects of use and disuse, with respect both to the body and mind."

Nor is this all. There is evidence that Mr. Darwin's belief in the efficiency of this factor, became stronger as he grew older and accumulated more evidence. The first of the extracts above given, taken from the sixth edition of the *Origin of Species*, runs thus:—

> "I think there can be no doubt that use in our domestic animals has strengthened and enlarged certain parts, and disuse diminished them; and that such modifications are inherited."

Now on turning to the first edition, p. 134, it will be found that instead of the words—"I think there can be no doubt," the words originally used were—"I think there can be *little* doubt." That this deliberate erasure of a qualifying word and substitution of a word implying unqualified belief, was due to a more decided recognition of a factor originally under-estimated, is clearly implied by the wording of the above-quoted passage from the preface to the *Descent of Man*; where he says that "*even* in the first edition of the 'Origin of Species,'" &c.: the implication being that much more in subsequent editions, and subsequent works, had he insisted on this factor. The change thus indicated is especially significant as having occurred at a time of life when the natural tendency is towards fixity of opinion.

During that earlier period when he was discovering the multitudinous cases in which his own hypothesis afforded solutions, and simultaneously observing how utterly futile in these multitudinous cases was the hypothesis propounded by his grandfather and Lamarck, Mr. Darwin was, not unnaturally, almost betrayed into the belief that the one is all-sufficient and the other inoperative. But in the mind of one so candid and ever open to more evidence, there naturally came a reaction. The inheritance of functionally-produced modifications, which, judging by the passage quoted above concerning the views of these earlier enquirers, would seem to have been at one time denied, but which as we have seen was always to some extent recognized, came to be recognized more and more, and deliberately included as a factor of importance.

Of this reaction displayed in the later writings of Mr. Darwin, let us now ask—Has it not to be carried further? Was the share in organic evolution which Mr. Darwin latterly assigned to the transmission of modifications caused by use and disuse, its due share? Consideration of the groups of evidences given above, will, I think, lead us to believe that its share has been much larger than he supposed even in his later days.

There is first the implication yielded by extensive classes of phenomena which remain inexplicable in the absence of this factor. If, as we see, co-operative parts do not vary together, even when few and close together, and may not therefore be assumed to do so when many and remote, we cannot account for those innumerable changes in organization which are implied when, for advantageous use of some modified part, many other parts which join it in action have to be modified.

Further, as increasing complexity of structure, accompanying increasing complexity of life, implies increasing number of faculties, of which each one conduces to preservation of self or descendants; and as the various individuals of a species, severally requiring something like the normal amounts of all these, may individually profit, here by an unusual amount of one, and there by an unusual amount of another; it follows that as the number of faculties becomes greater, it becomes more difficult for any one to be further developed by natural selection. Only where increase of some one is *predominantly* advantageous does the means seem adequate to the end. Especially in the case of powers which do not subserve self-preservation in appreciable degrees, does development by natural selection appear impracticable.

It is a fact recognized by Mr. Darwin, that where, by selection through successive generations, a part has been increased or decreased, its reaction upon other parts entails changes in them. This reaction is effected through the changes of function involved. If the changes of structure produced by such changes of function, are inheritable, then the re-adjustment of parts throughout the organism, taking place generation after generation, maintains an approximate balance; but if not, then generation after generation the organism must get more and more out of gear, and tend to become unworkable.

Further, as it is proved that change in the balance of functions registers its effects on the reproductive elements, we have to choose between the alternatives that the registered effects are irrelevant to the particular modifications which the organism has undergone, or that they are such as tend to produce repetitions of these modifications. The last of these alternatives makes the facts comprehensible; but the first of them not only leaves us with several unsolved problems, but is incongruous with the general truth that by reproduction, ancestral traits, down to minute details, are transmitted.

Though, in the absence of pecuniary interests and the interests in hobbies, no such special experiments as those which have established the inheritance of fortuitous variations have been made to ascertain whether functionally-produced modifications are inherited; yet certain apparent instances of such inheritance have forced themselves on observation without being sought for. In addition to other indications of a less conspicuous kind, is the one I have given above—the fact that the apparatus for tearing and mastication has decreased with decrease of its function, alike in civilized man and in some varieties of dogs which lead protected and pampered lives. Of the numerous cases named by Mr. Darwin, it is observable that they are yielded not by one class of parts only, but by most if not all classes—by the dermal system, the muscular system, the osseous system, the nervous system, the viscera; and that among parts liable to be functionally modified, the most numerous observed cases of inheritance are furnished by those which admit of preservation and easy comparison—the bones: these cases, moreover, being specially significant as showing how, in sundry unallied species, parallel changes of structure have occurred along with parallel changes of habit.

What, then, shall we say of the general implication? Are we to stop short with the admission that inheritance of functionally-produced modifications takes place only in cases in which there is evidence of it? May we properly assume that these many instances of changes of structure caused by changes of function, occurring in various tissues and various organs, are merely special and exceptional instances having no general significance? Shall we suppose that though the evidence which already exists has come to light without aid from a body of inquirers, there would be no great increase were due attention devoted to the collection of evidence? This is, I think, not a reasonable supposition. To me the *ensemble* of the facts suggests the belief, scarcely to be resisted, that the inheritance of functionally-produced modifications takes place universally. Looking at physiological phenomena as conforming to physical principles, it is difficult to conceive that a changed play of organic forces which in many cases of different kinds produces an inherited change of structure, does not do this in all cases. The implication, very strong I think, is that the action of every organ produces on it a reaction which, usually not altering its rate of nutrition, sometimes leaves it with diminished nutrition consequent on diminished action, and at other times increases its nutrition in proportion to its increased action; that while generating a modified *consensus* of functions and of structures, the activities are at the same time impressing this modified *consensus* on the sperm-cells and germ-cells whence future individuals are to be produced; and that in ways mostly too small to be identified, but occasionally in more conspicuous ways and in the course of generations, the resulting modifications of one or other kind show themselves. Further, it seems to me that as there are certain extensive classes of phenomena which are inexplicable if we assume the inheritance of fortuitous variations to be the sole factor, but which become at once explicable if we admit the inheritance of functionally-produced changes, we are justified in concluding that this inheritance of functionally-produced changes has been not simply a co-operating factor in organic evolution, but has been a co-operating factor without which organic evolution, in its higher forms at any rate, could never have taken place.

Be this or be it not a warrantable conclusion, there is, I think, good reason for a provisional acceptance of the hypothesis that the effects of use and disuse are inheritable; and for a methodic pursuit of inquiries with the view of either establishing it or disproving it. It seems scarcely reasonable to accept without clear demonstration, the belief that while a trivial difference of structure arising spontaneously is transmissible, a massive difference of structure, maintained generation after generation by change of function, leaves no trace in posterity. Considering that unquestionably the modification of structure by function is a *vera causa*, in so far as concerns the individual; and considering the number of facts which so competent an observer as Mr. Darwin regarded as evidence that transmission of such modifications takes place in particular cases; the hypothesis that such transmission takes place in conformity with a general law, holding of all active structures, should, I think, be regarded as at least a good working hypothesis.

But now supposing the broad conclusion above drawn to be granted—supposing all to agree that from the beginning, along with inheritance of useful variations fortuitously arising, there has been inheritance of effects produced by use and disuse; do there remain no classes of organic phenomena unaccounted for? To this question I think it must be replied that there do remain classes of organic phenomena unaccounted for. It may, I believe, be shown that certain cardinal traits of animals and plants at large are still unexplained; and that a further factor must be recognized. To show this, however, will require another paper.

II.

Ask a plumber who is repairing your pump, how the water is raised in it, and he replies—"By suction." Recalling the ability which he has to suck up water into his mouth through a tube, he is certain that he understands the pump's action. To inquire what he means by suction, seems to him absurd. He says you know as well as he does, what he means; and he cannot see that there is any need for asking how it happens that the water rises in the tube when he strains his mouth in a particular way. To the question why the pump, acting by suction, will not make the water rise above 32 feet, and practically not so much, he can give no answer; but this does not shake his confidence in his explanation.

On the other hand an inquirer who insists on knowing what suction is, may obtain from the physicist answers which give him clear ideas, not only about it but about many other things. He learns that on ourselves and all things around, there is an atmospheric pressure amounting to about 15 pounds on the square inch: 15 pounds being the average weight of a column of air having a square inch for its base and extending upwards from the sea-level to the limit of the Earth's atmosphere. He is made to observe that when he puts one end of a tube into water and the other end into his mouth, and then draws back his tongue, so leaving a vacant space, two things happen. One is that the pressure of air outside his cheeks, no longer balanced by an equal pressure of air inside, thrusts his cheeks inwards; and the other is that the pressure of air on the surface of the water, no longer balanced by an equal pressure of air within the tube and his mouth (into which part of the air from the tube has gone) the water is forced up the tube in consequence of the unequal pressure. Once understanding thus the nature of the so-called suction, he sees how it happens that when the plunger of the pump is raised and relieves from atmospheric pressure the water below it, the atmospheric pressure on the water in the well, not being balanced by that on the water in the tube, forces the water higher up the tube, so that it follows the plunger. And now he sees why the water cannot be raised beyond the theoretic limit of 32 feet: a limit made much lower in practice by imperfections in the apparatus. For if, simplifying the conception, he supposes the tube of the pump to be a square inch in section, then the atmospheric pressure of 15 pounds per square inch on the water in the well, can raise the water in the tube to such height only that the entire column of it weighs 15 pounds. Having been thus enlightened about the pump's action, the action of a barometer becomes intelligible. He perceives how, under the conditions established, the weight of the column of mercury balances that of an atmospheric column of equal diameter; and how, as the weight of the atmospheric column varies, there is a corresponding variation in the weight of the mercurial column,—shown by change of height. Moreover, having previously supposed that he understood the ascent of a balloon when he ascribed it to relative lightness, he now sees that he did not truly understand it. For he did not recognize it as a result of that upward pressure caused by the difference between the weight of the mass formed by the gas in the balloon *plus* the cylindrical column of air extending above it to the limit of the atmosphere, and the weight of a similar cylindrical column of air extending down to the under surface of the balloon: this difference of weight causing an equivalent upward pressure on the under surface.

Why do I introduce these familiar truths so entirely irrelevant to my subject? I do it to show, in the first place, the contrast between a vague conception of a cause and a distinct conception of it; or rather, the contrast between that conception of a cause which results when it is simply classed with some other or others which familiarity makes us think we understand, and that conception of a cause which results when it is represented in terms of definite physical forces admitting of measurement. And I do it to show, in the second place, that when we insist on resolving a verbally-intelligible cause into its actual factors, we get not only a clear solution of the problem before us, but we find that the way is opened to solutions of sundry other problems. While we rest satisfied with unanalyzed causes, we may be sure both that we do not rightly comprehend the production of the particular effects ascribed to them, and that we overlook other effects which would be revealed to us by contemplation of the causes as analyzed. Especially must this be so where the causation is complex. Hence we may infer that the phenomena presented by the development of species, are not likely to be truly conceived unless we keep in view the concrete agencies at work. Let us look closely at the facts to be dealt with.

The growth of a thing is effected by the joint operation of certain forces on certain materials; and when it dwindles, there is either a lack of some materials, or the forces co-operate in a way different from that which produces growth. If a structure has varied, the implication is that the processes which built it up were made unlike the parallel processes in other cases, by the greater or less amount of some one or more of the matters or actions concerned. Where there is unusual fertility, the play of vital activities is thereby shown to have deviated from the ordinary play of vital activities; and conversely, if there is infertility. If the germs, or ova, or seed, or offspring partially developed, survive more or survive less, it is either because their molar or molecular structures are unlike the average ones, or because they are affected in unlike ways by surrounding agencies. When life is prolonged, the fact implies that the combination of actions, visible and invisible, constituting life, retains its equilibrium longer than usual in presence of environing forces which tend to destroy its equilibrium. That is to say, growth, variation, survival, death, if they are to be reduced to the forms in which physical science can recognize them, must be expressed as effects of agencies definitely conceived—mechanical forces, light, heat, chemical affinity, &c.

This general conclusion brings with it the thought that the phrases employed in discussing organic evolution, though convenient and indeed needful, are liable to mislead us by veiling the actual agencies. That which really goes on in every organism is the working together of component parts in ways conducing to the continuance of their combined actions, in presence of things and actions outside; some of which tend to subserve, and others to destroy, the combination. The matters and forces in these two groups, are the sole causes properly so called. The words "natural selection," do not express a cause in the physical sense. They express a mode of co-operation among causes—or rather, to speak strictly, they express an effect of this mode of co-operation. The idea they convey seems perfectly intelligible. Natural selection having been compared with artificial selection, and the analogy pointed out, there apparently remains no indefiniteness: the inconvenience being, however, that the definiteness is of a wrong kind. The tacitly implied Nature which selects, is not an embodied agency analogous to the man who selects artificially; and the selection is not the picking out of an individual fixed on, but the overthrowing of many individuals by agencies which one successfully resists, and hence continues to live and multiply. Mr. Darwin was conscious of these misleading implications. In the introduction to his *Animals and Plants under Domestication* (p. 6) he says:—

> "For brevity sake I sometimes speak of natural selection as an intelligent power; ... I have, also, often personified the word Nature; for I have found it difficult to avoid this ambiguity; but I mean by nature only the aggregate action and product of many natural laws,—and by laws only the ascertained sequence of events."

But while he thus clearly saw, and distinctly asserted, that the factors of organic evolution are the concrete actions, inner and outer, to which every organism is subject, Mr. Darwin, by habitually using the convenient figure of speech, was, I think, prevented from recognizing so fully as he would otherwise have done, certain fundamental consequences of these actions.

Though it does not personalize the cause, and does not assimilate its mode of working to a human mode of working, kindred objections may be urged against the expression to which I was led when seeking to present the phenomena in literal terms rather than metaphorical terms—the survival of the fittest;[2] for in a vague way the first word, and in a clear way the second word, calls up an anthropocentric idea. The thought of survival inevitably suggests the human view of certain sets of phenomena, rather than that character which they have simply as groups of changes. If, asking what we really know of a plant, we exclude all the ideas associated with the words life and death, we find that the sole facts known to us are that there go on in the plant certain inter-dependent

---

[2] Though Mr. Darwin approved of this expression and occasionally employed it, he did not adopt it for general use; contending, very truly, that the expression Natural Selection is in some cases more convenient. See *Animals and Plants under Domestication* (first edition) Vol. i, p. 6; and *Origin of Species* (sixth edition) p. 49.

processes, in presence of certain aiding and hindering influences outside of it; and that in some cases a difference of structure or a favourable set of circumstances, allows these inter-dependent processes to go on for longer periods than in other cases. Again, in the working together of those many actions, internal and external, which determine the lives or deaths of organisms, we see nothing to which the words fitness and unfitness are applicable in the physical sense. If a key fits a lock, or a glove a hand, the relation of the things to one another is presentable to the perceptions. No approach to fitness of this kind is made by an organism which continues to live under certain conditions. Neither the organic structures themselves, nor their individual movements, nor those combined movements of certain among them which constitute conduct, are related in any analogous way to the things and actions in the environment. Evidently the word fittest, as thus used, is a figure of speech; suggesting the fact that amid surrounding actions, an organism characterized by the word has either a greater ability than others of its kind to maintain the equilibrium of its vital activities, or else has so much greater a power of multiplication that though not longer lived than they, it continues to live in posterity more persistently. And indeed, as we here see, the word fittest has to cover cases in which there may be less ability than usual to survive individually, but in which the defect is more than made good by higher degrees of fertility.

I have elaborated this criticism with the intention of emphasizing the need for studying the changes which have gone on, and are ever going on, in organic bodies, from an exclusively physical point of view. On contemplating the facts from this point of view, we become aware that, besides those special effects of the co-operating forces which eventuate in the longer survival of one individual than of others, and in the consequent increase through generations, of some trait which furthered its survival, many other effects are being wrought on each and all of the individuals. Bodies of every class and quality, inorganic as well as organic, are from instant to instant subject to the influences in their environments; are from instant to instant being changed by these in ways that are mostly inconspicuous; and are in course of time changed by them in conspicuous ways. Living things in common with dead things, are, I say, being thus perpetually acted upon and modified; and the changes hence resulting, constitute an all-important part of those undergone in the course of organic evolution. I do not mean to imply that changes of this class pass entirely unrecognized; for, as we shall see, Mr. Darwin takes cognizance of certain secondary and special ones. But the effects which are not taken into account, are those primary and universal effects which give certain fundamental characters to all organisms. Contemplation of an analogy will best prepare the way for appreciation of them, and of the relation they bear to those which at present monopolize attention.

An observant rambler along shores, will, here and there, note places where the sea has deposited things more or less similar, and separated them from dissimilar things—will see shingle parted from sand; larger stones sorted from smaller stones; and will occasionally discover deposits of shells more or less worn by being rolled about. Sometimes the pebbles or boulders composing the shingle at one end of a bay, he will find much larger than those at the other: intermediate sizes, having small average differences, occupying the space between the extremes. An example occurs, if I remember rightly, some mile or two to the west of Tenby; but the most remarkable and well-known example is that afforded by the Chesil bank. Here, along a shore some sixteen miles long, there is a gradual increase in the sizes of the stones; which, being at one end but mere pebbles, are at the other end immense boulders. In this case, then, the breakers and the undertow have effected a selection—have at each place left behind those stones which were too large to be readily moved, while taking away others small enough to be moved easily. But now, if we contemplate exclusively this selective action of the sea, we overlook certain important effects which the sea simultaneously works. While the stones have been differently acted upon in so far that some have been left here and some carried there; they have been similarly acted upon in two allied, but distinguishable, ways. By perpetually rolling them about and knocking them one against another, the waves have so broken off their most prominent parts as to produce in all of them more or less rounded forms; and then, further, the mutual friction of the stones simultaneously caused, has smoothed their surfaces. That is to say in general terms, the actions of environing agencies, so far as they have operated indiscriminately, have produced in the stones a certain unity of character; at the same time that they have, by their differential effects, separated them: the larger ones having withstood certain violent actions which the smaller ones could not withstand.

Similarly with other assemblages of objects which are alike in their primary traits but unlike in their secondary traits. When simultaneously exposed to the same set of actions, some of these actions, rising to a certain intensity, may be expected to work on particular members of the assemblage changes which they cannot work in those which are markedly unlike; while others of the actions will work in all of them similar changes, because of the uniform relations between these actions and certain attributes common to all members of the assemblage. Hence it is inferable that on living organisms, which form an assemblage of this kind, and are unceasingly exposed in common to the agencies composing their inorganic environments, there must be wrought two such sets of effects. There will result a universal likeness among them consequent on the likeness of their respective relations to the matters and forces around; and there will result, in some cases, the differences due to the differential effects of these matters and forces, and in other cases, the changes which, being life-sustaining or life-destroying, eventuate in certain natural selections.

I have, above, made a passing reference to the fact that Mr. Darwin did not fail to take account of some among these effects directly produced on organisms by surrounding inorganic agencies. Here are extracts from the sixth edition of the *Origin of Species* showing this.

> "It is very difficult to decide how far changed conditions, such as of climate, food, &c., have acted in a definite manner. There is reason to believe that in the course of time the effects have been greater than can be proved by clear evidence.... Mr.

Gould believes that birds of the same species are more brightly coloured under a clear atmosphere, than when living near the coast or on islands; and Wollaston is convinced that residence near the sea affects the colours of insects. Moquin-Tandon gives a list of plants which, when growing near the sea-shore, have their leaves in some degree fleshy, though not elsewhere fleshy" (pp. 106-7). "Some observers are convinced that a damp climate affects the growth of the hair, and that with the hair the horns are correlated" (p. 159).

In his subsequent work, *Animals and Plants under Domestication*, Mr. Darwin still more clearly recognizes these causes of change in organization. A chapter is devoted to the subject. After premising that "the direct action of the conditions of life, whether leading to definite or indefinite results, is a totally distinct consideration from the effects of natural selection;" he goes on to say that changed conditions of life "have acted so definitely and powerfully on the organisation of our domesticated productions, that they have sufficed to form new sub-varieties or races, without the aid of selection by man or of natural selection." Of his examples here are two.

"I have given in detail in the ninth chapter the most remarkable case known to me, namely, that in Germany several varieties of maize brought from the hotter parts of America were transformed in the course of only two or three generations." (Vol. ii, p. 277.) [And in this ninth chapter concerning these and other such instances he says "some of the foregoing differences would certainly be considered of specific value with plants in a state of nature." (Vol. i, p. 321.)] "Mr. Meehan, in a remarkable paper, compares twenty-nine kinds of American trees, belonging to various orders, with their nearest European allies, all grown in close proximity in the same garden and under as nearly as possible the same conditions." And then enumerating six traits in which the American forms all of them differ in like ways from their allied European forms, Mr. Darwin thinks there is no choice but to conclude that these "have been definitely caused by the long-continued action of the different climate of the two continents on the trees." (Vol. ii, pp. 281-2.)

But the fact we have to note is that while Mr. Darwin thus took account of special effects due to special amounts and combinations of agencies in the environment, he did not take account of the far more important effects due to the general and constant operation of these agencies.[3] If a difference between the quantities of a force which acts on two organisms, otherwise alike and otherwise similarly conditioned, produces some difference between them; then, by implication, this force produces in both of them effects which they show in common. The inequality between two things cannot have a value unless the things themselves have values. Similarly if, in two cases, some unlikeness of proportion among the surrounding inorganic agencies to which two plants or two animals are exposed, is followed by some

---

[3] It is true that while not deliberately admitted by Mr. Darwin, these effects are not denied by him. In his *Animals and Plants under Domestication* (vol. ii, 281), he refers to certain chapters in the *Principles of Biology*, in which I have discussed this general inter-action of the medium and the organism, and ascribed certain most general traits to it. But though, by his expressions, he implies a sympathetic attention to the argument, he does not in such way adopt the conclusion as to assign to this factor any share in the genesis of organic structures—much less that large share which I believe it has had. I did not myself at that time, nor indeed until quite recently, see how extensive and profound have been the influences on organization which, as we shall presently see, are traceable to the early results of this fundamental relation between organism and medium. I may add that it is in an essay on "Transcendental Physiology," first published in 1857, that the line of thought here followed out in its wider bearings, was first entered upon.

unlikeness in the changes wrought on them; then it follows that these several agencies taken separately, work changes in both of them. Hence we must infer that organisms have certain structural characters in common, which are consequent on the action of the medium in which they exist: using the word medium in a comprehensive sense, as including all physical forces falling upon them as well as matters bathing them. And we may conclude that from the primary characters thus produced there must result secondary characters.

Before going on to observe those general traits of organisms due to the general action of the inorganic environment upon them, I feel tempted to enlarge on the effects produced by each of the several matters and forces constituting the environment. I should like to do this not only to give a clear preliminary conception of the ways in which all organisms are affected by these universally-present agents, but also to show that, in the first place, these agents modify inorganic bodies as well as organic bodies, and that, in the second place, the organic are far more modifiable by them than the inorganic. But to avoid undue suspension of the argument, I content myself with saying that when the respective effects of gravitation, heat, light, &c, are studied, as well as the respective effects, physical and chemical, of the matters forming the media, water and air, it will be found that while more or less operative on all bodies, each modifies organic bodies to an extent immensely greater than the extent to which it modifies inorganic bodies.

Here, not discriminating among the special effects which these various forces and matters in the environment produce on both classes of bodies, let us consider their combined effects, and ask—What is the most general trait of such effects?

Obviously the most general trait is the greater amount of change wrought on the outer surface than on the inner mass. In so far as the matters of which the medium is composed come into play, the unavoidable implication is that they act more on the parts directly exposed to them than on the parts sheltered from them. And in so far as the forces pervading the medium come into play, it is manifest that, excluding gravity, which affects outer and inner parts indiscriminately, the outer parts have to bear larger shares of their actions. If it is a question of heat, then the exterior must lose it or gain it faster than the interior; and in a medium which is now warmer and now colder, the two must habitually differ in temperature to some extent—at least where the size is considerable. If it is a question of light, then in all but absolutely transparent masses, the outer parts must undergo more of any change producible by it than the inner parts—supposing other things equal; by which I mean, supposing the case is not complicated by any such convexities of the outer surface as produce internal concentrations of rays. Hence then, speaking generally, the necessity is that the primary and almost universal effect of the converse between the body and its medium, is to differentiate its outside from its inside. I say almost universal, because where the body is both mechanically and chemically stable, like, for instance, a quartz crystal, the medium may fail to work either inner or outer change.

Of illustrations among inorganic bodies, a convenient one is supplied by an old cannon-ball that has been long lying exposed. A coating of rust, formed of flakes within flakes, incloses it; and this thickens year by year, until, perhaps, it reaches a stage at which its exterior loses as much by rain and wind as its interior gains by further oxidation of the iron. Most mineral masses—pebbles, boulders, rocks—if they show any effect of the environment at all, show it only by that disintegration of surface which follows the freezing of absorbed water: an effect which, though mechanical rather than chemical, equally illustrates the general truth. Occasionally a "rocking-stone" is thus produced. There are formed successive layers relatively friable in texture, each of which, thickest at the most exposed parts, and being presently lost by weathering, leaves the contained mass in a shape more rounded than before; until, resting on its convex under-surface, it is easily moved. But of all instances perhaps the most remarkable is one to be seen on the west bank of the Nile at Philæ, where a ridge of granite 100 feet high, has had its outer parts reduced in course of time to a collection of boulder-shaped masses, varying from say a yard in diameter to six or eight feet, each one of which shows in progress an exfoliation of successively-formed shells of decomposed granite: most of the masses having portions of such shells partially detached.

If, now, inorganic masses, relatively so stable in composition, thus have their outer parts differentiated from their inner parts, what must we say of organic masses, characterized by such extreme chemical instability?—instability so great that their essential material is named protein, to indicate the readiness with which it passes from one isomeric form to another. Clearly the necessary inference is that this effect of the medium must be wrought inevitably and promptly, wherever the relation of outer and inner has become settled: a qualification for which the need will be seen hereafter.

Beginning with the earliest and most minute kinds of living things, we necessarily encounter difficulties in getting direct evidence; since, of the countless species now existing, all have been subject during millions upon millions of years to the evolutionary process, and have had their primary traits complicated and obscured by those endless secondary traits which the natural selection of favourable variations has produced. Among protophytes it needs but to think of the multitudinous varieties of diatoms and desmids, with their elaborately-constructed coverings; or of the definite methods of growth and multiplication among such simple *Algæ* as the *Conjugatæ*; to see that most of their distinctive characters are due to inherited constitutions, which have been slowly moulded by survival of the fittest to this or that mode of life. To disentangle such parts of their developmental changes as are due to the action of the medium, is therefore hardly possible. We can hope only to get a general conception of it by contemplating the totality of the facts.

The first cardinal fact is that all protophytes are cellular—all show us this contrast between outside and inside. Supposing the multitudinous specialities of the envelope in different orders and genera of protophytes to be set against one another, and mutually cancelled, there remains as a trait common to them—an envelope unlike that which it envelopes. The second cardinal fact is that this simple trait is the earliest trait displayed in germs, or spores, or other parts from which new individuals are to arise; and that, consequently, this trait must be regarded as having been primordial. For it is an established truth of organic evolution that embryos show us, in general ways, the forms of remote ancestors; and that the first changes undergone, indicate, more or less clearly, the first changes which took place in the series of forms through which the existing form has been reached. Describing, in successive groups of plants, the early transformations of these primitive units, Sachs[4] says of the lowest Algæ that "the conjugated protoplasmic body clothes itself with a cell-wall" (p. 10); that in "the spores of Mosses and Vascular Cryptogams" and in "the pollen of Phanerogams" ... "the protoplasmic body of the mother-cell breaks up into four lumps, which quickly round themselves off and contract, and become enveloped by a cell-membrane only after complete separation" (p. 13); that in the *Equisetaceæ* "the young spores, when first separated, are still naked, but they soon become surrounded by a cell-membrane" (p. 14); and that in higher plants, as in the pollen of many Dicotyledons,

---

[4] *Text-Book of Botany*, &c. by Julius Sachs. Translated by A. W. Bennett and W. T. T. Dyer.

"the contracting daughter-cells secrete cellulose even during their separation" (p. 14). Here, then, in whatever way we interpret it, the fact is that there quickly arises an outer layer different from the contained matter. But the most significant evidence is furnished by "the masses of protoplasm that escape into water from the injured sacs of *Vaucheria*, which often instantly become rounded into globular bodies," and of which the "hyaline protoplasm envelopes the whole as a skin" (p. 41) which "is denser than the inner and more watery substance" (p. 42). As in this case the protoplasm is but a fragment, and as it is removed from the influence of the parent-cell, this differentiating process can scarcely be regarded as anything more than the effect of physico-chemical actions: a conclusion which is supported by the statement of Sachs that "not only every vacuole in a solid protoplasmic body, but also every thread of protoplasm which penetrates the sap-cavity, and finally the inner side of the protoplasm-sac which encloses the sap-cavity, is also bounded by a skin" (p. 42). If then "every portion of a protoplasmic body immediately surrounds itself, when it becomes isolated, with such a skin," which is shown in all cases to arise at the surface of contact with sap or water, this primary differentiation of outer from inner must be ascribed to the direct action of the medium. Whether the coating thus initiated is secreted by the protoplasm, or whether, as seems more likely, it results from transformation of it, matters not to the argument. Either way the action of the medium causes its formation; and either way the many varied and complex differentiations which developed cell-walls display, must be considered as originating from those variations of this physically-generated covering which natural selection has taken advantage of.

The contained protoplasm of a vegetal cell, which has self-mobility and when liberated sometimes performs amœba-like motions for a time, may be regarded as an imprisoned amœba; and when we pass from it to a free amœba, which is one of the simplest types of first animals, or *Protozoa*, we naturally meet with kindred phenomena. The general trait which here concerns us, is that while its plastic or semi-fluid sarcode goes on protruding, in irregular ways, now this and now that part of its periphery, and again withdrawing into its interior first one and then another of these temporary processes, perhaps with some small portion of food attached, there is but an indistinct differentiation of outer from inner (a fact shown by the frequent coalescence of the pseudopodia in Rhizopods); but that when it eventually becomes quiescent, the surface becomes differentiated from the contents: the passing into an encysted state, doubtless in large measure due to inherited proclivity, being furthered, and having probably been once initiated, by the action of the medium. The connexion between constancy of relative position among the parts of the sarcode, and the rise of a contrast between superficial and central parts, is perhaps best shown in the minutest and simplest *Infusoria*, the *Monadinæ*. The genus *Monas* is described by Kent as "plastic and unstable in form, possessing no distinct cuticular investment; ... the food-substances incepted at all parts of the periphery";[5] and the genus *Scytomonas* he says "differs from *Monas* only in its persistent shape and accompanying greater rigidity of the peripheral or ectoplasmic layer."[6] Describing generally such low forms, some of which are said to have neither nucleus nor vacuole, he remarks that in types somewhat higher "the outer or peripheral border of the protoplasmic mass, while not assuming the character of a distinct cell-wall or so-called cuticle, presents, as compared with the inner substance of that mass, a slightly more solid type of composition."[7] And it is added that these forms having so slightly differentiated an exterior, "while usually exhibiting a more or less characteristic normal outline, can revert at will to a pseud-amœboid and repent state."[8] Here, then, we have several

---

[5] *A Manual of the Infusoria*, by W. Saville Kent. Vol. i, p. 232.
[6] *Ib.* Vol. i, p. 241.
[7] Kent, Vol. i, p. 56.
[8] *Ib.* Vol. i, p. 57.

indications of the truth that the permanent externality of a certain part of the substance, is followed by transformation of it into a coating unlike the substance it contains. Indefinite and structureless in the simplest of these forms, as instance again the *Gregarina*,[9] the limiting membrane becomes, in higher *Infusoria*, definite and often complex: showing that the selection of favourable variations has had largely to do with its formation. In such types as the *Foraminifera*, which, almost structureless internally though they are, secrete calcareous shells, it is clear that the nature of this outer layer is determined by inherited constitution. But recognition of this consists with the belief that the action of the medium initiated the outer layer, specialized though it now is; and that even still, contact with the medium excites secretion of it.

A remarkable analogy remains to be named. When we study the action of the medium in an inorganic mass, we are led to see that between the outer changed layer and the inner unchanged mass, comes a surface where active change is going on. Here we have to note that, alike in the plant-cell and in the animal-cell, there is a similar relation of parts. Immediately inside the envelope comes the primordial utricle in the one case, and in the other case the layer of active sarcode. In

---

[9] *The Elements of Comparative Anatomy*, by T. H. Huxley, pp. 7-9.

either case the living protoplasm, placed in the position of a lining to the cuticle of the cell, is shielded from the direct action of the medium, and yet is not beyond the reach of its influences.

Limited, as thus far drawn, to a certain common trait of those minute organisms which are mostly below the reach of unaided vision, the foregoing conclusion appears trivial enough. But it ceases to appear trivial on passing into a wider field, and observing the implications, direct and indirect, as they concern plants and animals of sensible sizes.

Popular expositions of science have so far familiarized many readers with a certain fundamental trait of living things around, that they have ceased to perceive how marvellous a trait it is, and, until interpreted by the Theory of Evolution, how utterly mysterious. In past times, the conception of an ordinary plant or animal which prevailed, not throughout the world at large only but among the most instructed, was that it is a single continuous entity. One of these living things was unhesitatingly regarded as being in all respects a unit. Parts it might have, various in their sizes, forms, and compositions; but these were components of a whole which had been from the beginning in its original nature a whole. Even to naturalists fifty years ago, the assertion that a cabbage or a cow, though in one sense a whole, is in another sense a vast society of minute individuals, severally living in greater or less degrees, and some of them maintaining their independent lives unrestrained, would have seemed an absurdity. But this truth which, like so many of the truths established by science, is contrary to that common sense in which most people have so much confidence, has been gradually growing clear since the days when Leeuwenhoek and his contemporaries began to examine through lenses the minute structures of common plants and animals. Each improvement in the microscope, while it has widened our knowledge of those minute forms of life described above, has revealed further evidence of the fact that all the larger forms of life consist of units severally allied in their fundamental traits to these minute forms of life. Though, as formulated by Schwann and Schleiden, the cell-doctrine has undergone qualifications of statement; yet the qualifications have not been such as to militate against the general proposition that organisms visible to the naked eye, are severally compounded of invisible organisms—using that word in its most comprehensive sense. And then, when the development of any animal is traced, it is found that having been primarily a nucleated cell, and having afterwards become by spontaneous fission a cluster of nucleated cells, it goes on through successive stages to form out of such cells, ever multiplying and modifying in various ways, the several tissues and organs composing the adult.

On the hypothesis of evolution this universal trait has to be accepted not as a fact that is strange but unmeaning. It has to be accepted as evidence that all the visible forms of life have arisen by union of the invisible forms; which, instead of flying apart when they divided, remained together. Various intermediate stages are known. Among plants, those of the *Volvox* type show us the component protophytes so feebly combined that they severally carry on their lives with no appreciable subordination to the life of the group. And among animals, a parallel relation between the lives of the units and the life of the group is shown us in *Uroglena* and *Syncrypta*. From these first stages upwards, may be traced through successively higher types, an increasing subordination of the units to the aggregate; though still a subordination leaving to them conspicuous amounts of individual activity. Joining which facts with the phenomena presented by the cell-multiplication and aggregation of every unfolding germ, naturalists are now accepting the conclusion that by this process of composition from *Protozoa*, were formed all classes of the *Metazoa*[10]—(as animals formed by this compounding are now called); and that in a similar way from *Protophyta*, were formed all classes of what I suppose will be called *Metaphyta*, though the word does not yet seem to have become current.

And now what is the general meaning of these truths, taken in connexion with the conclusion reached in the last section. It is that this universal trait of the *Metazoa* and *Metaphyta*, must be ascribed to the primitive action and re-action between the organism and its medium. The operation of those forces which produced the primary differentiation of outer from inner in early minute masses of protoplasm, pre-determined this universal cell-structure of all embryos, plant and animal, and the consequent cell-composition of adult forms arising from them. How unavoidable is this implication, will be seen on carrying further an illustration already used—that of the shingle-covered shore, the pebbles on which, while being in some cases selected, have been in all cases rounded and smoothed. Suppose a bed of such shingle to be, as we often see it, solidified, along with interfused material, into a conglomerate. What in such case must be considered as the chief trait of such

---

[10] *A Treatise on Comparative Embryology*, by F. M. Balfour, Vol. ii, chap. xiii.

conglomerate; or rather—what must we regard as the chief cause of its distinctive characters? Evidently the action of the sea. Without the breakers, no pebbles; without the pebbles, no conglomerate. Similarly then, in the absence of that action of the medium by which was effected the differentiation of outer from inner in those microscopic portions of protoplasm constituting the earliest and simplest animals and plants, there could not have existed this cardinal trait of composition which all the higher animals and plants show us.

So that, active as has been the part played by natural selection, alike in modifying and moulding the original units—largely as survival of the fittest has been instrumental in furthering and controlling the combination of these units into visible organisms, and eventually into large ones; yet we must ascribe to the direct effect of the medium on the first forms of life, that character of which this everywhere-operative factor has taken advantage.

Let us turn now to another and more obvious attribute of higher organisms, for which also there is this same general cause. Let us observe how, on a higher platform, there recurs this differentiation of outer from inner—how this primary trait in the living units with which life commences, re-appears as a primary trait in those aggregates of such units which constitute visible organisms.

In its simplest and most unmistakable form, we see this in the early changes of an unfolding ovum of primitive type. The original fertilized single cell, having by spontaneous fission multiplied into a cluster of such cells, there begins to show itself a contrast between periphery and centre; and presently there is formed a sphere consisting of a superficial layer unlike its contents. The first change, then, is the rise of a difference between that outer part which holds direct converse with the surrounding medium, and that inclosed part which does not. This primary differentiation in these compound embryos of higher animals, parallels the primary differentiation undergone by the simplest living things.

Leaving, for the present, succeeding changes of the compound embryo, the significance of which we shall have to consider by-and-by, let us pass now to the adult forms of visible plants and animals. In them we find cardinal traits which, after what we have seen above, will further impress us with the importance of the effects wrought on the organism by its medium.

From the thallus of a sea-weed up to the leaf of a highly developed phænogam, we find, at all stages, a contrast between the inner and outer parts of these flattened masses of tissue. In the higher *Algæ* "the outermost layers consist of smaller and firmer cells, while the inner cells are often very large, and sometimes extremely long;"[11] and in the leaves of trees the epidermal layer, besides differing in the sizes and shapes of its component cells from the parenchyma forming the inner substance of the leaf, is itself differentiated by having a continuous cuticle, and by having the outer walls of its cells unlike the inner walls.[12] Especially significant is the structure of such intermediate types as the Liverworts. Beyond the differentiation of the covering cells from the contained cells, and the contrast between upper surface and under surface, the frond of *Marchantia polymorpha* clearly shows us the direct effect of incident forces; and shows us, too, how it is involved with the effect of inherited proclivities. The frond grows from a flat disc-shaped gemma, the two sides of which are alike. Either side may fall uppermost; and then of the developing shoot, the side exposed to the light "is under all circumstances the upper side which forms stomata, the dark side becomes the under side which produces root-hairs and leafy processes."[13] So that while we have undeniable proof that the contrasted influences of the medium on the two sides,

---

[11] Sachs, p. 210.
[12] *Ibid.* pp. 83-4.
[13] *Ibid.* p. 185.

initiate the differentiation, we have also proof that the completion of it is determined by the transmitted structure of the type; since it is impossible to ascribe the development of stomata to the direct action of air and light. On turning from foliar expansions, to stems and roots, facts of like meaning meet us. Speaking generally of epidermal tissue and inner tissue, Sachs remarks that "the contrast of the two is the plainer the more the part of the plant concerned is exposed to air and light."[14] Elsewhere, in correspondence with this, it is said that in roots the cells of the epidermis, though distinguished by bearing hairs, "are otherwise similar to those of the fundamental tissue" which they clothe,[15] while the cuticular covering is relatively thin; whereas in stems the epidermis (often further differentiated) is composed of layers of cells which are smaller and thicker-walled: a stronger contrast of structure corresponding to a stronger contrast of conditions. By way of meeting the suggestion that these respective differences are wholly due to the natural selection of favourable variations, it will suffice if I draw attention to the unlikeness between imbedded roots and exposed roots. While in darkness, and surrounded by moist earth, the outermost protective coats, even of large roots, are comparatively

---

[14] *Ibid.* p. 80.
[15] Sachs, p. 83.

thin; but when the accidents of growth entail permanent exposure to light and air, roots acquire coverings allied in character to the coverings of branches. That the action of the medium causes these and converse changes, cannot be doubted when we find, on the one hand, that "roots can become directly transformed into leaf-bearing shoots," and, on the other hand, that in some plants certain "apparent roots are only underground shoots," and that nevertheless "they are similar to true roots in function and in the formation of tissue, but have no root-cap, and, when they come to the light above ground, continue to grow in the manner of ordinary leaf-shoots."[16] If, then, in highly developed plants inheriting pronounced structures, this differentiating influence of the medium is so marked, it must have been all-important at the outset while types were undetermined.

As with plants so with animals, we find good reason for inferring that while the specialities of the tegumentary parts must be ascribed to the natural selection of favourable variations, their most general traits are due to the direct action of surrounding agencies. Here we come upon the border of those changes which are ascribable to use and disuse. But from this class of changes we may fitly exclude those

---

[16] *Ibid.* p. 147.

in which the parts concerned are wholly or mainly passive. A corn and a blister will conveniently serve to illustrate the way in which certain outer actions initiate in the superficial tissues, effects of very marked kinds, which are related neither to the needs of the organism nor to its normal structure. They are neither adaptive changes nor changes towards completion of the type. After noting them we may pass to allied, but still more instructive, changes. Continuous pressure on any portion of the surface causes absorption, while intermittent pressure causes growth: the one impeding circulation and the passage of plasma from the capillaries into the tissues, and the other aiding both. There are yet further mechanically-produced effects. That the general character of the ribbed skin on the under surfaces of the feet and insides of the hands is directly due to friction and intermittent pressure, we have the proofs:—first, that the tracts most exposed to rough usage are the most ribbed; second, that the insides of hands subject to unusual amounts of rough usage, as those of sailors, are strongly ribbed all over; and third, that in hands which are very little used, the parts commonly ribbed become quite smooth. These several kinds of evidence, however, full of meaning as they are, I give simply to prepare the way for evidence of a much more conclusive kind.

Where a wide ulcer has eaten away the deep-seated layer out of which the epidermis grows, or where this layer has been destroyed by an extensive burn, the process of healing is very significant. From the subjacent tissues, which in the normal order have no concern with outward growth, there is produced a new skin, or rather a pro-skin; for this substituted outward-growing layer contains no hair-follicles or other specialities of the original one. Nevertheless, it is like the original one in so far that it is a continually renewed protective covering. Doubtless it may be contended that this make-shift skin results from the inherited proclivity of the type—the tendency to complete afresh the structure of the species when injured. We cannot, however, ignore the immediate influence of the medium, on recalling the facts above named, or on remembering the further fact that an inflamed surface of skin, when not sheltered from the air, will throw out a film of coagulable lymph. But that the direct action of the medium is a chief factor we are clearly shown by another case. Accident or disease occasionally causes permanent eversion, or protrusion, of mucous membrane. After a period of irritability, great at first but decreasing as the change advances, this membrane assumes the general character of ordinary skin. Nor is this all: its microscopic structure changes. Where it is a mucous membrane of the kind covered by cylinder-epithelium, the cylinders gradually shorten, becoming finally flat, and there results a squamous epithelium: there is a near approach in minute composition to epidermis. Here a tendency towards completion of the type cannot be alleged; for there is, contrariwise, divergence from the type. The effect of the medium is so great that, in a short time, it overcomes the inherited proclivity and produces a structure of opposite kind to the normal one.

With but little break we come here upon a significant analogy, parallel to an analogy already described. As was pointed out, an inorganic body that is modifiable by its medium, acquires, after a time, an outer coat which has already undergone such change as surrounding agencies can effect; has a contained mass which is as yet unchanged, because unreached; and has a surface between the two where change is going on—a region of activity. And we saw that alike in the vegetal cell and the animal cell there exist analogous distributions: of course with the difference that the innermost part is not inert. Now we have to note that in those aggregates of cells constituting the *Metaphyta* and *Metazoa*, analogous distributions also exist. In plants they are of course not to be looked for in leaves and other deciduous portions, but only in portions of long duration—stems and branches. Naturally, too, we need not expect them in plants having modes of growth which early produce an outer practically dead part, that effectually shields the inner actively living part of the stem from the influence of the medium—long-lived acrogens such as tree-ferns and long-lived endogens such as palms. But in the highest plants, exogens, which have the actively living part of their stems within reach of environing agencies, we find this part,—the cambium layer,—is one from which there is a growth inwards forming wood, and a growth outwards forming bark: there is an increasingly thick covering (where it does not scale off) of tissue changed by the medium, and inside this a film of highest vitality. In so far as concerns the present argument, it is the same with the *Metazoa*, or at least all of them which have developed organizations. The outer skin grows up from a limiting plane, or layer, a little distance below the surface—a place of predominant vital activity. Here perpetually arise new cells, which, as they develop, are thrust outwards and form the epidermis: flattening and drying up as they approach the surface, whence, having for a time served to shield the parts below, they finally scale off and leave younger ones to take their places. This still undifferentiated tissue forming the base of the epidermis, and existing also as a source of renewal in internal organs, is the essentially living substance; and facts above given imply that it was the action of the medium on this essentially living substance, which, during early stages in the organization of the *Metazoa*, initiated that protective envelope which presently became an inherited structure—a structure which, though now mainly inherited, still continues to be modifiable by its initiator.

Fully to perceive the way in which these evidences compel us to recognize the influence of the medium as a primordial factor, we need but conceive them as interpreted without it. Suppose, for instance, we say that the structure of the epidermis is wholly determined by the natural selection of favourable variations; what must be the position taken in presence of the fact above named, that when mucous membrane is exposed to the air its cell-structure changes into the cell-structure of skin? The position taken must be this:—Though mucous membrane in a highly-evolved individual organism, thus shows the powerful effect of the medium on its surface; yet we must not suppose that the medium had the effect of producing such a cell-structure on the surfaces of primitive forms, undifferentiated though they were; or, if we suppose that such an effect was produced on them, we must not suppose that it was inheritable. Contrariwise, we must suppose that such effect of the medium either was not wrought at all, or that it was evanescent: though repeated through millions upon millions of generations it left no traces. And we must conclude that this skin-structure arose only in consequence of spontaneous variations not physically initiated (though like those physically initiated) which natural selection laid hold of and increased. Does any one think this a tenable position?

And now we approach the last and chief series of morphological phenomena which must be ascribed to the direct action of environing matters and forces. These are presented to us when we study the early stages in the development of the embryos of the *Metazoa* in general.

We will set out with the fact already noted in passing, that after repeated spontaneous fissions have changed the original fertilized germ-cell into that cluster of cells which forms a gemmule or a primitive ovum, the first contrast which arises is between the peripheral parts and the central parts. Where, as with lower creatures which do not lay up large stores of nutriment with the germs of their offspring, the inner mass is inconsiderable, the outer layer of cells, which are presently made quite small by repeated subdivisions, forms a membrane extending over the whole surface—the blastoderm. The next stage of development, which ends in this covering layer becoming double, is reached in two ways—by invagination and by delamination; but which is the original way and which the abridged way, is not quite certain. Of invagination, multitudinously exemplified in the lowest types, Mr. Balfour says:—"On purely *à priori* grounds there is in my opinion more to be said for invagination than for any other view";[17] and, for present purposes, it will suffice if we limit ourselves to this: making its nature clear to the general reader by a simple illustration.

Take a small india-rubber ball—not of the inflated kind, nor of the solid kind, but of the kind about an inch or so in diameter with a small hole through which, under pressure, the air escapes. Suppose that instead of consisting of india-rubber its wall consists of small cells made polyhedral in form by mutual pressure, and united together. This will represent the blastoderm. Now with the finger, thrust in one side of the ball until it touches the other: so making a cup. This action will stand for the process of invagination. Imagine that by continuance of it, the hemispherical cup becomes very much deepened and the opening narrowed, until the cup becomes a sac, of which the introverted wall is everywhere in contact with the outer wall. This will represent the two-layered

---

[17] *A Treatise on Comparative Embryology.* By Francis M. Balfour, LL.D., F.R.S. Vol. ii, p. 343 (second edition).

"gastrula"—the simplest ancestral form of the *Metazoa*: a form which is permanently represented in some of the lowest types; for it needs but tentacles round the mouth of the sac, to produce a common hydra. Here the fact which it chiefly concerns us to remark, is that of these two layers the outer, called in embryological language the epiblast, continues to carry on direct converse with the forces and matters in the environment; while the inner, called the hypoblast, comes in contact with such only of these matters as are put into the food-cavity which it lines. We have further to note that in the embryos of *Metazoa* at all advanced in organization, there arises between these two layers a third—the mesoblast. The origin of this is seen in types where the developmental process is not obscured by the presence of a large food-yolk. While the above-described introversion is taking place, and before the inner surfaces of the resulting epiblast and hypoblast have come into contact, cells, or amœboid units equivalent to them, are budded off from one or both of these inner surfaces, or some part of one or other; and these form a layer which eventually lies between the other two—a layer which, as this mode of formation implies, never has any converse with the surrounding medium and its contents, or with the nutritive bodies taken in from it. The striking facts to which this description is a necessary introduction, may now be stated. From the outer layer, or epiblast, are developed the permanent epidermis and its out-growths, the nervous system, and the organs of sense. From the introverted layer, or hypoblast, are developed the alimentary canal and those parts of its appended organs, liver, pancreas, &c., which are concerned in delivering their secretions into the alimentary canal, as well as the linings of those ramifying tubes in the lungs which convey air to the places where gaseous exchange is effected. And from the mesoblast originate the bones, the muscles, the heart and blood-vessels, and the lymphatics, together with such parts of various internal organs as are most remotely concerned with the outer world. Minor qualifications being admitted, there remain the broad general facts, that out of that part of the external layer which remains permanently external, are developed all the structures which carry on intercourse with the medium and its contents, active and passive; out of the introverted part of this external layer, are developed the structures which carry on intercourse with the quasi-external substances that are taken into the interior—solid food, water, and air; while out of the mesoblast are developed structures which have never had, from first to last, any intercourse with the environment. Let us contemplate these general facts.

Who would have imagined that the nervous system is a modified portion of the primitive epidermis? In the absence of proofs furnished by the concurrent testimony of embryologists during the last thirty or forty years, who would have believed that the brain arises from an unfolded tract of the outer skin, which, sinking down beneath the surface, becomes imbedded in other tissues and eventually surrounded by a bony case? Yet the human nervous system in common with the nervous systems of lower animals is thus originated. In the words of Mr. Balfour, early embryological changes imply that—

> "the functions of the central nervous system, which were originally taken by the whole skin, became gradually concentrated in a special part of the skin which was step by step removed from the surface, and has finally become in the higher types a well-defined organ imbedded in the subdermal tissues.... The embryological evidence shows that the ganglion-cells of the central part of the nervous system are originally derived from the simple undifferentiated epithelial cells of the surface of the body."[18]

Less startling perhaps, though still startling enough, is the fact that the eye is evolved out of a portion of the

---

[18] Balfour, l.c. Vol. ii, 400-1.

skin; and that while the crystalline lens and its surroundings thus originate, the "percipient portions of the organs of special sense, especially of optic organs, are often formed from the same part of the primitive epidermis" which forms the central nervous system.[19] Similarly is it with the organs for smelling and hearing. These, too, begin as sacs formed by infoldings of the epidermis; and while their parts are developing they are joined from within by nervous structures which were themselves epidermic in origin. How are we to interpret these strange transformations? Observing, as we pass, how absurd from the point of view of the special-creationist, would appear such a filiation of structures, and such a round-about mode of embryonic development, we have here to remark that the process is not one to have been anticipated as a result of natural selection. After numbers of spontaneous variations had occurred, as the hypothesis implies, in useless ways, the variation which primarily initiated a nervous centre might reasonably have been expected to occur in some internal part where it would be fitly located. Its initiation in a dangerous place and subsequent migration to a safe place, would be incomprehensible. Not so if we bear in mind the cardinal truth above set forth, that the structures for holding converse with the

---

[19] Balfour, l.c. Vol. ii, p. 401.

medium and its contents, arise in that completely superficial part which is directly affected by the medium and its contents; and if we draw the inference that the external actions themselves initiate the structures. These once commenced, and furthered by natural selection where favourable to life, would form the first term of a series ending in developed sense organs and a developed nervous system.[20]

Though it would enforce the argument, I must, for brevity's sake, pass over the analogous evolution of that introverted layer, or hypoblast, out of which the alimentary canal and attached organs arise. It will suffice to emphasize the fact that having been originally external, this layer continues in its developed form to have a quasi-externality, alike in its digesting part and in its respiratory part; since it continues to deal with matters alien to the organism. I must also refrain from dwelling at length on the fact already adverted to, that the intermediate derived layer, or mesoblast, which was at the outset completely internal, originates those structures which ever remain completely internal, and have no communication with the environment save through the structures developed from the other two: an antithesis which has great significance.

---

[20] For a general delineation of the changes by which the development is effected, see Balfour, l.c. Vol. ii, pp. 401-4.

Here, instead of dwelling on these details, it will be better to draw attention to the most general aspect of the facts. Whatever may be the course of subsequent changes, the first change is the formation of a superficial layer or blastoderm; and by whatever series of transformations the adult structure is reached, it is from the blastoderm that all the organs forming the adult originate. Why this marvellous fact?

Meaning is given to it if we go back to the first stage in which *Protozoa*, having by repeated fissions formed a cluster, then arranged themselves into a hollow sphere, as do the protophytes forming a *Volvox*. Originally alike all over its surface, the hollow sphere of ciliated units thus formed, would, if not quite spherical, assume a constant attitude when moving through the water; and hence one part of the spheroid would more frequently than the rest come in contact with nutritive matters to be taken in. A division of labour resulting from such a variation being advantageous, and tending therefore to increase in descendants, would end in a differentiation like that shown in the gemmules of various low types of *Metazoa*, which, ovate in shape, are ciliated over one part of the surface only. There would arise a form in which the cilium-bearing units effected locomotion and aeration; while on the others, assuming an amœba-like character, devolved the function of absorbing food: a primordial specialization variously indicated by evidence.[21] Just noting that an ancestral origin of this kind is implied by the fact that in low types of *Metazoa* a hollow sphere of cells is the form first assumed by the unfolding embryo, I draw attention to the point here of chief interest; namely that the primary differentiation of this hollow sphere is in such case determined by a difference in the converse of its parts with the medium and its contents; and that the subsequent invagination arises by a continuance of this differential converse.

Even neglecting this first stage and commencing with the next, in which a "gastrula" has been produced by the permanent introversion of one portion of the surface of the hollow sphere, it will suffice if we consider what must thereafter have happened. That which continued to be the outer surface was the part which from time to time touched quiescent masses and occasionally received the collisions consequent on its own motions or the motions of other things. It was the part to receive the sound-vibrations occasionally propagated through the water; the part to be affected more strongly than any other by those variations in the amounts of light caused by the passing of small bodies close to it; and the part which met those diffused molecules constituting odours. That is to say,

---

[21] *See* Balfour, Vol. i, 149 and Vol. ii, 313-4.

from the beginning the surface was the part on which there fell the various influences pervading the environment, the part by which there was received those impressions from the environment serving for the guidance of actions, and the part which had to bear the mechanical re-actions consequent upon such actions. Necessarily, therefore, the surface was the part in which were initiated the various instrumentalities for carrying on intercourse with the environment. To suppose otherwise is to suppose that such instrumentalities arose internally where they could neither be operated on by surrounding agencies nor operate on them,—where the differentiating forces did not come into play, and the differentiated structures had nothing to do; and it is to suppose that meanwhile the parts directly exposed to the differentiating forces remained unchanged. Clearly, then, organization could not but begin on the surface; and having thus begun, its subsequent course could not but be determined by its superficial origin. And hence these remarkable facts showing us that individual evolution is accomplished by successive in-foldings and in-growings. Doubtless natural selection soon came into action, as, for example, in the removal of the rudimentary nervous centres from the surface; since an individual in which they were a little more deeply seated would be less likely to be incapacitated by injury of them. And so in multitudinous other ways. But nevertheless, as we here see, natural selection could operate only under subjection. It could do no more than take advantage of those structural changes which the medium and its contents initiated.

See, then, how large has been the part played by this primordial factor. Had it done no more than give to *Protozoa* and *Protophyta* that cell-form which characterizes them—had it done no more than entail the cellular composition which is so remarkable a trait of *Metazoa* and *Metaphyta*—had it done no more than cause the repetition in all visible animals and plants of that primary differentiation of outer from inner which it first wrought in animals and plants invisible to the naked eye; it would have done much towards giving to organisms of all kinds certain leading traits. But it has done more than this. By causing the first differentiations of those clusters of units out of which visible animals in general arose, it fixed the starting place for organization, and therefore determined the course of organization; and, doing this, gave indelible traits to embryonic transformations and to adult structures.

Though mainly carried on after the inductive method, the argument at the close of the foregoing section has passed into the deductive. Here let us follow for a space the deductive method pure and simple. Doubtless in biology *à priori* reasoning is dangerous; but there can be no danger in considering whether its results coincide with those reached by reasoning *à posteriori*.

Biologists in general agree that in the present state of the world, no such thing happens as the rise of a living creature out of non-living matter. They do not deny, however, that at a remote period in the past, when the temperature of the Earth's surface was much higher than at present, and other physical conditions were unlike those we know, inorganic matter, through successive complications, gave origin to organic matter. So many substances once supposed to belong exclusively to living bodies, have now been formed artificially, that men of science scarcely question the conclusion that there are conditions under which, by yet another step of composition, quaternary compounds of lower types pass into those of highest types. That there once took place gradual divergence of the organic from the inorganic, is, indeed, a necessary implication of the hypothesis of Evolution, taken as a whole; and if we accept it as a whole, we must put to ourselves the question—What were the early stages of progress which followed, after the most complex form of matter had arisen out of forms of matter a degree less complex?

At first, protoplasm could have had no proclivities to one or other arrangement of parts; unless, indeed, a purely mechanical proclivity towards a spherical form when suspended in a liquid. At the outset it must have been passive. In respect of its passivity, primitive organic matter must have been like inorganic matter. No such thing as spontaneous variation could have occurred in it; for variation implies some habitual course of change from which it is a divergence, and is therefore excluded where there is no habitual course of change. In the absence of that cyclical series of metamorphoses which even the simplest living thing now shows us, as a result of its inherited constitution, there could be no *point d'appui* for natural selection. How, then, did organic evolution begin?

If a primitive mass of organic matter was like a mass of inorganic matter in respect of its passivity, and differed only in respect of its greater changeableness; then we must infer that its first changes conformed to the same general law as do the changes of an inorganic mass. The instability of the homogeneous is a universal principle. In all cases the homogeneous tends to pass into the heterogeneous, and the less heterogeneous into the more heterogeneous. In the primordial units of protoplasm, then, the step with which evolution commenced must have been the passage from a state of complete likeness throughout the mass to a state in which there existed some unlikeness. Further, the cause of this step in one of these portions of organic matter, as in any portion of inorganic matter, must have been the different exposure of its parts to incident forces. What incident forces? Those of its medium or environment. Which were the parts thus differently exposed? Necessarily the outside and the inside. Inevitably, then, alike in the organic aggregate and the inorganic aggregate (supposing it to have coherence enough to maintain constant relative positions among its parts), the first fall from homogeneity to heterogeneity must always have been the differentiation of the external surface from the internal contents. No matter whether the modification was physical or chemical, one of composition or of decomposition, it comes within the same generalization. The direct action of the medium was the primordial factor of organic evolution.

And now, finally, let us look at the factors in their *ensemble*, and consider the respective parts they play: observing, especially, the ways in which, at successive stages, they severally give place one to another in degree of importance.

Acting alone, the primordial factor must have initiated the primary differentiation in all units of protoplasm alike. I say alike, but I must forthwith qualify the word. For since surrounding influences, physical and chemical, could not be absolutely the same in all places, especially when the first rudiments of living things had spread over a considerable area, there necessarily arose small contrasts between the degrees and kinds of superficial differentiation effected. As soon as these became decided, natural selection came into play; for inevitably the unlikenesses produced among the units had effects on their lives: there was survival of some among the modified forms rather than others. Utterly in the dark though we are respecting the causes which set up that process of fission everywhere occurring among the minutest forms of life, we must infer that, when established, it furthered the spread of those which were most favourably differentiated by the medium. Though natural selection must have become increasingly active when once it had got a start; yet the differentiating action of the medium never ceased to be a co-operator in the development of these first animals and plants. Again taking the lead as there arose the composite forms of animals and plants, and again losing the lead with that advancing differentiation of these higher types which gave more scope to natural selection, it nevertheless continued, and must ever continue, to be a cause, both direct and indirect, of modifications in structure.

Along with that remarkable process which, beginning in minute forms with what is called conjugation, developed into sexual generation, there came into play causes of frequent and marked fortuitous variations. The mixtures of constitutional proclivities made more or less unlike by unlikenesses of physical conditions, inevitably led to occasional concurrences of forces producing deviations of structure. These were of course mostly suppressed, but sometimes increased, by survival of the fittest. When, along with the growing multiplication in forms of life, conflict and competition became continually more active, fortuitous variations of structure of no account in the converse with the medium, became of much account in the struggle with enemies and competitors; and natural selection of such variations became the predominant factor. Especially throughout the plant-world its action appears to have been immensely the most important; and throughout that large part of the animal world characterized by relative inactivity, the survival of individuals that had varied in favourable ways, must all along have been the chief cause of the divergence of species and the occasional production of higher ones.

But gradually with that increase of activity which we see on ascending to successively higher grades of animals, and especially with that increased complexity of life which we also see, there came more and more into play as a factor, the inheritance of those modifications of structure caused by modifications of function. Eventually, among creatures of high organization, this factor became an important one; and I think there is reason to conclude that, in the case of the highest of creatures, civilized men, among whom the kinds of variation which affect survival are too multitudinous to permit easy selection of any one, and among whom survival of the fittest is greatly interfered with, it has become the chief factor: such aid as survival of the fittest gives, being usually limited to the preservation of those in whom the totality of the faculties has been most favourably moulded by functional changes.

Of course this sketch of the relations among the factors must be taken as in large measure a speculation. We are now too far removed from the beginnings of life to obtain data for anything more than tentative conclusions respecting its earliest stages; especially in the absence of any clue to the mode in which multiplication, first agamogenetic and then gamogenetic, was initiated. But it has seemed to me not amiss to present this general conception, by way of showing how the deductive interpretation harmonizes with the several inferences reached by induction.

In his article on Evolution in the *Encyclopædia Britannica,* Professor Huxley writes as follows:—

> "How far 'natural selection' suffices for the production of species remains to be seen. Few can doubt that, if not the whole cause, it is a very important factor in that operation.... On the evidence of palaeontology, the evolution of many existing forms of animal life from their predecessors is no longer an hypothesis, but an historical fact; it is only the nature of the physiological factors to which that evolution is due which is still open to discussion."

With these passages I may fitly join a remark made in the admirable address Prof. Huxley delivered before unveiling the statue of Mr. Darwin in the Museum at South Kensington. Deprecating the supposition that an authoritative sanction was given by the ceremony to the current ideas concerning organic evolution, he said that "science commits suicide when it adopts a creed."

Along with larger motives, one motive which has joined in prompting the foregoing articles, has been the desire to point out that already among biologists, the beliefs concerning the origin of species have assumed too much the character of a creed; and that while becoming settled they have been narrowed. So far from further broadening that broader view which Mr. Darwin reached as he grew older, his followers appear to have retrograded towards a more restricted view than he ever expressed. Thus there seems occasion for recognizing the warning uttered by Prof. Huxley, as not uncalled for.

Whatever may be thought of the arguments and conclusions set forth in this article and the preceding one, they will perhaps serve to show that it is as yet far too soon to close the inquiry concerning the causes of organic evolution.

NOTE.

After the above articles were published, I received from Dr. Downes a copy of a paper "On the Influence of Light on Protoplasm," written by himself and Mr. T. P. Blunt, M.A., which was communicated to the Royal Society in 1878. It was a continuation of a preceding paper which, referring chiefly to *Bacteria*, contended that—

> "Light is inimical to, and under favourable conditions may wholly prevent, the development of these organisms."

This supplementary paper goes on to show that the injurious effect of light upon protoplasm results only in presence of oxygen. Taking first a comparatively simple type of molecule which enters into the composition of organic matter, the authors say, after detailing experiments:—

> "It was evident, therefore, that oxygen was the agent of destruction under the influence of sunlight."

And accounts of experiments upon minute organisms are followed by the sentence—

> "It seemed, therefore, that in absence of an atmosphere, light failed entirely to produce

any effect on such organisms as were able to appear."

They sum up the results of their experiments in the paragraph—

"We conclude, therefore, both from analogy and from direct experiment, that the observed action on these organisms is not dependent on light per se, but that the presence of free oxygen is necessary; light and oxygen together accomplishing what neither can do alone: and the inference seems irresistible that the effect produced is a gradual oxidation of the constituent protoplasm of these organisms, and that, in this respect, protoplasm, although living, is not exempt from laws which appear to govern the relations of light and oxygen to forms of matter less highly endowed. A force which is indirectly absolutely essential to life as we know it, and matter in the absence of which life has not yet been proved to exist, here unite for its destruction."

What is the obvious implication? If oxygen in presence of light destroys one of these minutest portions of protoplasm, what will be its effect on a larger portion of protoplasm? It will work an effect on the surface instead of on the whole mass. Not like the minutest mass made inert all through, the larger mass will be made inert only on its outside; and, indeed, the like will happen with the minutest mass if the light or the oxygen is very small in quantity. Hence there will result an envelope of changed matter, inclosing and protecting the unchanged protoplasm—there will result a rudimentary cell-wall.

www.ingramcontent.com/pod-product-compliance
Lightning Source LLC
Chambersburg PA
CBHW071200240526
45470CB00017B/617